T0220400

Control Grid Motion Estimation for Efficient Application of Optical Flow

Synthesis Lectures on Algorithms and Software in Engineering

Editor
Andreas Spanias, *Arizona State University*

Control Grid Motion Estimation for Efficient Application of Optical Flow
Christine M. Zwart and David H. Frakes
2013

Sparse Representations for Radar with MATLAB® Examples
Peter Knee
2012

Analysis of the MPEG-1 Layer III (MP3) Algorithm Using MATLAB
Jayaraman J. Thiagarajan and Andreas Spanias
2011

Theory and Applications of Gaussian Quadrature Methods
Narayan Kovvali
2011

Algorithms and Software for Predictive and Perceptual Modeling of Speech
Venkatraman Atti
2011

Adaptive High-Resolution Sensor Waveform Design for Tracking
Ioannis Kyriakides, Darryl Morrell, and Antonia Papandreou-Suppappola
2010

MATLAB® Software for the Code Excited Linear Prediction Algorithm: The Federal
Standard-1016
Karthikeyan N. Ramamurthy and Andreas S. Spanias
2010

OFDM Systems for Wireless Communications
Adarsh B. Narasimhamurthy, Mahesh K. Banavar, and Cihan Tepedelenliouglu
2010

Advances in Modern Blind Signal Separation Algorithms: Theory and Applications
Kostas Kokkinakis and Philipos C. Loizou
2010

Advances in Waveform-Agile Sensing for Tracking
Sandeep Prasad Sira, Antonia Papandreou-Suppappola, and Darryl Morrell
2008

Despeckle Filtering Algorithms and Software for Ultrasound Imaging
Christos P. Loizou and Constantinos S. Pattichis
2008

Control Grid Motion Estimation for Efficient Application of Optical Flow
Christine M. Zwart and David H. Frakes

ISBN: 978-3-031-00392-9   paperback
ISBN: 978-3-031-01520-5   ebook

DOI 10.1007/978-3-031-01520-5

A Publication in the Springer series
SYNTHESIS LECTURES ON ALGORITHMS AND SOFTWARE IN ENGINEERING

Lecture #11
Series Editor: Andreas Spanias, *Arizona State University*
Series ISSN
Synthesis Lectures on Algorithms and Software in Engineering
Print 1938-1727   Electronic 1938-1735

Control Grid Motion Estimation for Efficient Application of Optical Flow

Christine M. Zwart and David H. Frakes

Ira A. Fulton Schools of Engineering, Arizona State University

SYNTHESIS LECTURES ON ALGORITHMS AND SOFTWARE #11

ABSTRACT

Motion estimation is a long-standing cornerstone of image and video processing. Most notably, motion estimation serves as the foundation for many of today's ubiquitous video coding standards including H.264. Motion estimators also play key roles in countless other applications that serve the consumer, industrial, biomedical, and military sectors. Of the many available motion estimation techniques, optical flow is widely regarded as most flexible. The flexibility offered by optical flow is particularly useful for complex registration and interpolation problems, but comes at a considerable computational expense. As the volume and dimensionality of data that motion estimators are applied to continue to grow, that expense becomes more and more costly. Control grid motion estimators based on optical flow can accomplish motion estimation with flexibility similar to pure optical flow, but at a fraction of the computational expense. Control grid methods also offer the added benefit of representing motion far more compactly than pure optical flow. This booklet explores control grid motion estimation and provides implementations of the approach that apply to data of multiple dimensionalities. Important current applications of control grid methods including registration and interpolation are also developed.

KEYWORDS

motion estimation, image registration, interpolation, optimization methods, piecewise linear techniques

Contents

CHAPTER 1

Introduction

Motion is the process by which the position of an object varies with time. Motion estimation in video processing is generally based on temporal registration of video frames or images. Techniques for registration and motion estimation are widely available, and methods exist for diverse combinations of flexibility, resolution, and assumptions. Method selection involves trade-offs in accuracy, sensitivity, and efficiency. In this chapter, we briefly review the considerations, applications, and history of motion estimation techniques for video registration before exploring two popular methods (block-matching and optical flow). At the end of the chapter, we briefly cover conventions used in the remainder of the book and preview its organization.

1.1   REGISTRATION AND MOTION ESTIMATION

Registration is the process of defining the spatial transformation that best maps an image or model to another image or model. A broad range of methods exists for identifying the transformation and assessing the correspondence or effectiveness of the results. A general expression for a geometric transform is:

$$\mathbf{H}(\hat{x}, \hat{y}) = \mathbf{H}(a(x, y), b(x, y)) = \mathbf{I}(x, y), \tag{1.1}$$

where $\mathbf{H}$ is the transformed or output image and $\mathbf{I}$ is the input image. The geometric transform specified by the functions $a(x, y)$ and $b(x, y)$ describes the relationship between the two coordinate frames such that data initially located at some location (x, y) in the input are repositioned and found at location $(\hat{x}, \hat{y})$ in the output. In some applications, the transform may be an analytical function of x and y. Alternatively, a piece- or even pixel-wise definition of the new coordinates may be required.

For motion estimation, it is common to express the transformation in terms of a displacement field:

$$a(x, y) = x + \mathbf{D_x}(x, y), \tag{1.2}$$
$$b(x, y) = y + \mathbf{D_y}(x, y), \tag{1.3}$$

where the elements of $\mathbf{D_x}$ and $\mathbf{D_y}$ describe the displacement in the x and y dimensions, respectively. The freedom and detail with which the vector field is specified determines the flexibility of the motion model.

Video registration and motion estimation look to identify the motion field that generates the approximated frame $\hat{\mathbf{I}}(x, y, t)$ that best corresponds to the true frame $\mathbf{I}(x, y, t)$. The video frame collected at time t is transformed into an estimate of the frame at time $t + 1$ as:

$$\hat{\mathbf{I}}(x, y, t + 1) = \mathbf{I}(x + \mathbf{D_x}(x, y), y + \mathbf{D_y}(x, y), t). \tag{1.4}$$

The procedures and metrics for determining the 'best' motion field are determined by the requirements of a given application.

Motion estimation applications can be classified as motion-final and motion-supported. For motion-final applications, the estimated motion field is the final output of the process. Further quantification or interrogation of the motion vectors may follow, but the purpose of the motion estimation step is to construct a depiction of the motion. Applications involving estimated motion fields are highly varied and appear in current research in fields as diverse as fluid mechanics [Shindler et al., 2012], civil engineering [Rodriguez et al., 2012], robotics [Arkin, 2012], astronomy [Fujita et al., 2012], aeronautics [Shabayek et al., 2011], medicine [Li et al., 2002], and agriculture [Dawkins et al., 2012]. In contrast, motion-supported applications utilize motion vectors as inputs to other processing steps. Video compression is perhaps the most pervasive motion-supported application in modern video processing [Wiegand et al., 2003]. Other applications include motion-directed interpolation [Chen and Lorenz, 2012] and deinterlacing [Dufaux and Moscheni, 1995].

Depending on the application, the desired features of the motion estimation algorithm can vary greatly. For motion-final applications, high accuracy and resolution may be important while computation time may be less critical [Brox et al., 2010]. In applications requiring a dense and accurate depiction of the motion field, methods based on optical flow are commonly utilized [Amiaz et al., 2007]. For motion-supported applications, the computation time and feasibility of implementation in basic hardware may be most important [Wójcikowski et al., 2011]. Additional considerations like the smoothness or entropy of the motion field may influence what constitutes the 'best' result [Jing et al., 2003]. For time sensitive and hardware implemented applications, block-based motion estimation algorithms are standard [Chatterjee and Chakrabarti, 2011].

While a tremendous number of other approaches and applications exist, in this book we will restrict our coverage of background methods to block matching and optical flow. In Chapter 2 we will show that control grid methods represent a compromise resulting in motion field density, accuracy, and computational costs intermediate to block matching and optical flow algorithms. In addition to being the most relevant methods for prefacing our discussion of control grid techniques, block matching and optical flow are the most popular brute force and mathematical optimization methods (respectively) for motion estimation. In Chapters 3 and 4, we will highlight applications of control grid methods to problems beyond motion estimation. Other approaches to those applications will be covered at that time.

1.2   BLOCK-BASED MOTION ESTIMATION

Block-based motion estimation describes the class of methods that assign a displacement value pair $\mathbf{d} = [d_x, d_y]$ to blocks of pixels such that the motion vector describing the movement of the block applies to all pixels within the block. Block-based methods are frequently referred to as block-matching algorithms (BMAs) as the objective function is phrased in terms of similarity between the originating block and blocks of pixels in the target image or frame. BMAs define the block-based motion field by identifying the displacement vector that shifts the original block of pixels into a position where it overlays the most similar block in the target frame. Algorithms are differentiated by the sequence through which candidate positions are considered, limits and assumptions about which candidates are possible or probable given previous matches or image features, and the metric by which similarity is assessed [Purwar et al., 2011].

The most common metrics for assessing match quality are the sum of absolute differences:

$$SAD = \sum_{m=1}^{k} \sum_{n=1}^{k} |\mathbf{H}(m, n) - \mathbf{I}(m, n)|, \tag{1.5}$$

and sum of squared differences:

$$SSD = \sum_{m=1}^{k} \sum_{n=1}^{k} [\mathbf{H}(m, n) - \mathbf{I}(m, n)]^2. \tag{1.6}$$

In both cases the comparison occurs pixel-wise, and the summation is over the full block where k indicates the block side length and $\mathbf{H}$ and $\mathbf{I}$ are the blocks being compared. Many other metrics have been proposed including metrics based on the number of 'close' pixel matches and cross correlation [Choi and Jeong, 2011]. To reduce the computational burden, some approaches evaluate match quality based on partial blocks, projections, or histograms of the blocks rather than on a pixel-by-pixel basis over the full block [Park et al., 2012]. Alternatively, the Fourier transform can be employed to streamline the computation of the difference metric [Kilthau et al., 2002].

The most basic approach to block matching is an exhaustive search. Exhaustive searching is generally limited to some maximum displacement range and can be conducted with sub-pixel resolution via interpolation. Alternative approaches to conventional block matching involve intelligent methods for eliminating portions of the exhaustive search either by pre-discarding candidate displacements or avoiding or simplifying calculation of similarity metrics [Song and Akoglu, 2011]. Since the exhaustive search is algorithmically straightforward and foundational to all BMAs, we will briefly present a simple software implementation.

Program 1.1: Implementation of Exhaustive Search Block Matching Algorithm

```
1   function [d1,...%displacments along vertical
2             d2]...%displacments along horizontal
3             = BMA_ES(originImage,...%source image
4                     destinationImage,...%image with ...
                        matching candidates
5                     blockSize,...%size of the matching blocks
6                     maxDisplacement)%largest displacement ...
                        to be considered
7
8   %Computes displacement field using exhaustive search method
9
10  [M N] = size(originImage); %sizing
11
```

```
12  d2 = zeros(M,N); %preallocate
13  d1 = zeros(M,N); %preallocate
14  %Find the best match for each blockSize x blockSize block ...
        beginning with the
15  %block having pixel (1,1) as its top left corner pixel.
16  %outer loops index through the blocks in the origin image
17  for m = 1 : blockSize : M−blockSize+1
18      for n = 1 : blockSize : N−blockSize+1
19          %inner loops shift through match candidates in the ...
                destination image
20          %prep with a non−shift output:
21          blockDelta = abs(...
22              originImage(m:m+(blockSize −1),...
23                      n:n+(blockSize −1)) − ...
24              destinationImage(m:m+(blockSize −1),...
25                      n:n+(blockSize −1)));
26          minError = sum(blockDelta(:));%using sum of abs. differences
27          d1Best = 0; %populate the current estimate
28          d2Best = 0; %for the displacement field entry
29
30          for d1Candidate = −maxDisplacement : maxDisplacement
31              for d2Candidate = −maxDisplacement : maxDisplacement
32                  %the candidate block in the destination image has
33                  %the top left corner:
34                  dest_m = m + d1Candidate;
35                  dest_n = n + d2Candidate;
36
37                  %make sure the destination block is entirely ...
                        within bounds
38                  if ( dest_m < 1 ||... %top
39                      dest_n < 1 ||... %left
40                      dest_m + blockSize −1 > M ||... %bottom
41                      dest_n + blockSize −1 > N ) %right
42                          continue; % do not consider block
43                  end
44                  %compute the absolute difference in the blocks:
45                  blockDelta = abs(...
46                   originImage(m:m+(blockSize −1),...
47                          n:n+(blockSize −1)) − ...
48                   destinationImage((dest_m:dest_m+(blockSize −1)),...
49                          (dest_n:dest_n+(blockSize −1))));
50                  SAD = sum(blockDelta(:));%using sum of abs. ...
                        differences
51                  if(SAD<minError)
52                      minError = SAD;
53                      d1Best = d1Candidate; %populate the current ...
                            estimate
54                      d2Best = d2Candidate; %for the displacement ...
                            field entry
55                  end
56
57              end
58          end
59
60          %assign the displacements that emerge as the best to the ...
                block:
```

```
61              d1(m:m+(blockSize−1),n:n+(blockSize−1)) = d1Best;
62              d2(m:m+(blockSize−1),n:n+(blockSize−1)) = d2Best;
63        end
64    end
```

In the block matching approach described in Program 1.2, several key assumptions are made:

1. Each block of pixels moves as a unit.

2. The blocks do not move by more than the maximum displacement.

3. All possible displacements must be checked. Equivalently, each of the possible displacement vectors is equally likely.

All BMAs are subject to the first assumption and almost all (for computational feasibility) make some assertion regarding the maximum displacement. At a minimum, most BMAs assert that the matched block exists in the adjacent frame. Most BMAs also make some assumption regarding the probability of the displacement candidates. The exhaustive search approach always computes the match quality achieved by each possible displacement within the search range. From a probabilistic perspective, the only reason to do an exhaustive search would be that all of the possible displacements are equally probable. Other assumptions or causes for an exhaustive search include coding (particularly hardware) considerations and an unwillingness to tolerate missing lower error matches regardless of how improbable they may be.

When coding challenges can be overcome and some increases in error are tolerable, additional assumptions can be applied to adjust and build on Assumption 3. Common assumptions involve similarity of the displacement field to a previously computed set, smoothness of the error surface, and predominance of smaller displacements [Saha et al., 2011]. As a result, efficient algorithms tend to begin by searching a central region and narrowly expanding the search region based on the preliminary error or beginning with a less sensitive, lower resolution approach before narrowing in on the final full resolution displacement estimate [Chun and Ra, 1994]. Alternately, the initial search region may begin around a displaced location using the displacement field previously computed for a different frame pair [Shi et al., 2011]. Rapid BMAs are commonly referenced by the search pattern or number of steps involved. For example, spiral search, diamond search, three step search, new three step search, and four step search are all modern BMAs.

In addition to algorithmic developments focused on novel search progressions and maximizing the quality of the block matching, recent efforts are frequently focused on approaches that effectively decrease memory and computational burden in ways that translate well to implementation in hardware [Song and Akoglu, 2011]. Hardware focused research is generally motivated by video compression [Wiegand et al., 2003]. Some of these efforts are focused on the implementation of alternative search trajectories. Other approaches still involve checking all matches but with metrics requiring fewer computations or truncated computations allowing a poor match to be ruled out early based on a partial comparison [Lin et al., 2012].

While the computational burden associated with the exhaustive search block matching algorithm can be minimized using intelligent speed-ups, several fundamental limitations related to resolution and flexibility remain. The output of any BMA is one motion vector per block of pixels. This affords motion models capable of describing the translation of block-sized objects. While specific applications like video compression are designed around a block-based model, the low resolution poses significant challenges to realistically modeling three-dimensional motion. However, as motion models are allowed to take on more complex forms and are resolved at finer scales, trial-and-error approaches to match optimization become unwieldy. Optical flow, covered next, is one of the most common approaches to generating dense displacement fields capable of describing more complex motion as a combination of pixel-by-pixel translations [Horn and Schunck, 1981, Lucas and Kanade, 1981].

1.3   OPTICAL FLOW

Optical or optic flow describes a class of motion estimation approaches based on what is typically referred to as the brightness constraint. Approximating optical flow amounts to matching sensor responses over time; for digital video frames, this amounts to matching pixels with the same intensity. For a given pixel-sized object (point light source) located at spatial location (x, y) at time t, optical flow looks to identify the object's new location, $(x + \alpha, y + \beta)$, at time $t + \Delta$. Tying the object's identity to its measured intensity, $\mathbf{I}$, this matching amounts to the following equality:

$$\mathbf{I}(x, y, t) = \mathbf{I}(x + \alpha, y + \beta, t + \Delta). \tag{1.7}$$

Equation 1.7 is the known as the brightness constraint. Optical flow uses brightness-based matching to construct a two-dimensional projection of three-dimensional motion.

Strictly speaking, block matching algorithms represent an approach to optical flow [Little and Verri, 1989]. By matching block intensities across frames, BMAs empirically determine the displacements that best satisfy Equation 1.7 for a given block of pixels. Other methods for determining the optimal displacement values have been introduced including phase, frequency, and causality derived approaches [Yamashita et al., 2012]. In this section, we will focus on the class of optical flow algorithms that utilize first order derivatives to determine the optimal displacements.

Classical approaches to optical flow begin with a Taylor series expansion of the brightness constraint (Equation 1.7):

$$\mathbf{I}(x, y, t) = \mathbf{I}(x, y, t) + \alpha \frac{\partial \mathbf{I}(x, y, t)}{\partial x} + \beta \frac{\partial \mathbf{I}(x, y, t)}{\partial y} + \Delta \frac{\partial \mathbf{I}(x, y, t)}{\partial t} + H.O.T. \tag{1.8}$$

Higher order terms are generally dropped (some authors have retained second order terms [Nagel, 1983]) and the partial derivates are approximated from the digital data. Using the first order Taylor series expansion and $\mathbf{I_x}, \mathbf{I_y}$, and $\mathbf{I_t}$ to represent the approximations to the x, y, and t partial derivatives, the deviation from the brightness constraint (squared error) can be rewritten as:

$$E(\alpha, \beta) = \left(\alpha \mathbf{I_x}(x, y, t) + \beta \mathbf{I_y}(x, y, t) + \Delta \mathbf{I_t}(x, y, t) \right)^2. \tag{1.9}$$

Typically, the spacing in time is fixed at $\Delta = 1$ and the displacements are normalized such that:

$$[\mathbf{D_x}, \mathbf{D_y}] = \underset{[\mathbf{D_x}, \mathbf{D_y}]}{\mathrm{argmin}} \left(\mathbf{D_x} \circ \mathbf{I_x} + \mathbf{D_y} \circ \mathbf{I_y} + \mathbf{I_t} \right)^2, \tag{1.10}$$

where $[\mathbf{D_x}, \mathbf{D_y}]$ is the vector field describing the estimated motion or 'optical flow' and $\circ$ indicates the Hadamard product (element-by-element multiplication). For any given pixel location, Equation 1.9 may not define a unique displacement pair $[\mathbf{D_x}(x, y), \mathbf{D_y}(x, y)]$. This is commonly described as the aperture problem and is highlighted in the underdetermined equation:

$$0 = \left(\mathbf{D_x}(x, y, t)\mathbf{I_x}(x, y, t) + \mathbf{D_y}(x, y, t)\mathbf{I_y}(x, y, t) + \mathbf{I_t}(x, y, t) \right)^2, \tag{1.11}$$

where $\mathbf{D_x}(x, y, t)$ and $\mathbf{D_x}(x, y, t)$ are both unknowns. An optical flow algorithm must specify an additional constraint to overcome the aperture problem. Typical constraints relate to the smoothness of the motion field either globally, in the region local to the pixel in question, or a hybrid [Bruhn et al., 2005]. In this book we will explore control grid optimization. Control grids provide a hybrid smoothness constraint that is locally smooth and maintains global connectivity and continuity. Details of this type of constraint are explored in Chapter 2 along with the control grid approach to optical flow.

1.4   CONVENTIONS

Throughout this chapter we have introduced several conventions for describing motion estimation problems mathematically. Several of these conventions are worth formalizing as we will continue to employ them throughout the book. We also take this time to introduce other key conventions that will be used in subsequent chapters.

In general, we will use boldface, capital letters to represent matrices (e.g., $\mathbf{I}$), boldface lower case letters to represent vectors (e.g., $\mathbf{d}$), and standard type lower case letters to represent scalars (e.g., x). Values or matrices that are modified or approximated versions of existing values or data are indicated with a hat (e.g., $\hat{I}$). In subsequent chapters, we will use a bar to indicate matrices that have been vectorized retaining the capital and boldface convention (e.g., $\bar{I}$).

Figure 1.1 depicts the conventions used in describing positions within images and indexing matrices. We will use x to indicate position along the continuous vertical axis and y to indicate position along the continuous horizontal axis. The coordinate pair (x, y) can be used to describe any position within the image. To refer to pixels within the $M \times N$ image matrix, we use (m, n) with m and n both integers. When indexing, we use a one-based system such that the top left corner is has coordinates (m, n). This combination of standards maps directly to indexing by row and column numbers in array notation (e.g., $\mathbf{I}_{m,n}$) and common mathematical programming languages like MATLAB. We do not use a separate indexing notation for displaying and storing images.

1.5   ORGANIZATION OF THE BOOK

The remainder of this book focuses on describing the methods and applications of registration transforms based on Control Grid Interpolation (CGI). CGI is a hybridization of block- and pixel-

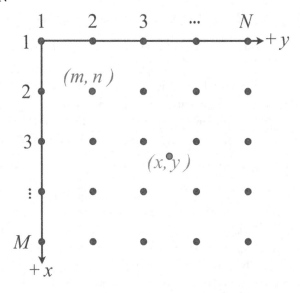

Figure 1.1: The $M \times N$ grid of pixels is indexed as (m, n). m values are the integer index points along the x axis and correspond to row entries in the data matrix. n values are integer indexes along the y access and correspond to column entries.

based registration techniques (e.g., block-matching and optical flow) with direct applications to problems related to motion (spatial displacement over time). The CGI approach to registration can also be used to link similar data features in non-motion contexts. In the second chapter, the algorithm's conventional formulation is reviewed along with several more recently developed approaches based on the same foundational mathematics. These developments relate to the selection of block size and the optimization mathematics. The third and fourth chapters cover applications of CGI. Problems related directly to registration are covered first followed by applications related to interpolation. Final conclusions are presented in the fifth chapter.

CHAPTER 2

Control Grid Interpolation (CGI)

The terms control grid and control grid interpolation as they are used in this book stem from the geometric registration algorithms packaged with the Video Image Communication And Retrieval (VICAR) image processing software system developed by Jet Propulsion Laboratories (JPL) [Castleman, 1979]. As originally presented, a control grid is a mesh of contiguous quadrilaterals, the vertices of which are mapped to the vertices of a regular lattice of contiguous rectangles as shown in Figure 2.1. Control grid interpolation describes the process by which points within the elements of the control grid are mapped between the input and output images.

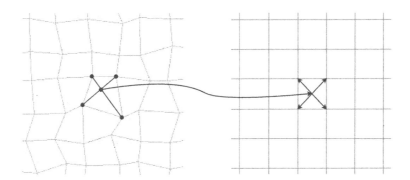

Figure 2.1: Illustration of the control grid and control grid interpolation as defined in Castleman [1979]. Note that the vertices are mapped directly and that the relative positions within grid elements are maintained following the transform.

Sullivan and Baker introduced control grid interpolation (CGI) for motion estimation in 1991. We use their modified definition for control grid interpolation:

> Control grid interpolation is a technique for performing spatial image transformations. It begins with specified spatial displacements for a small number of points in an image, termed control points. The spatial displacements of all other points in the image are then determined by interpolating between the control point displacements [Sullivan and Baker, 1991].

This definition removes the requirement that the output grid be a regular lattice and emphasizes the description of the transform in terms of displacements. Fundamentally, any spatial transform or motion field explicitly defined at control points or nodes and interpolated in between meets the modified definition. While the use of quadrilateral (rectangular and otherwise) elements is retained by Sullivan and Baker [1991], their proposed definition applies to a wide range of spatial transformation-based approaches to motion estimation including those that utilize alternative meshing structures such as triangles and multiresolution girds [Huang and Hsu, 1994, Nakaya and Harashima, 1994].

Also unrestricted in this definition of control grid interpolation is the method for determining the displacements at the nodes. Whereas Sullivan and Baker implemented CGI motion estimation as an iterative refinement of displacements computed using block matching techniques, other authors have smoothed and approximated dense motion fields computed using optical flow with CGI. Still other approaches utilize full or partial control grid connectivity to overcome the aperture problem and constrain optical flow-based approaches to motion estimation [Altunbasak and Tekalp, 1997]. In this book we focus on the latter, utilizing CGI as a regularization framework for solving optimization problems related to the brightness constraint as it applies to optical flow and other image processing applications.

In this chapter, we introduce the mathematics and some of the basic algorithms associated with approaching CGI as a constraint. CGI will be used to reduce the number of unknowns in an optimization problem from the number of points in the data set to the number of nodes in the control grid. While all of the applications discussed in subsequent chapters will use error functions related to the brightness constraint, that restriction is not necessary in exploring the preliminary interpolation mathematics of CGI. We will cover the basic interpolation foundation in one and two dimensions restricting the scope of our discussion to linear interpolation (bilinear for two dimensions) and fixed-size rectangular grid elements. We then explore modified control grid structures with multiresolution grids and adaptive sizing of grid elements. Following our introduction of the framework, we explore the specifics of the optical flow optimization mathematics as well as approaches for minimizing optical flow cost functions analytically.

2.1 CONVENTIONAL CGI FORMULATION

We begin describing the basic interpolation mathematics of CGI in terms of uniformly spaced control points starting with a simplified, one-dimensional format followed by a two-dimensional, rectangular lattice.

2.1.1 ONE-DIMENSIONAL

In one dimension, the control grid structure reduces to a series of adjacent segments. Rather than break an array into sub-elements, the nodes divide a vector into chunks of data. We restrict the scope of our discussion and assert that values between nodes are interpolated linearly. Higher order interpolation approaches are also possible and have been explored by other authors [Szeliski and Coughlan, 1994].

For values of the vector $\mathbf{d}$ defined at adjacent nodes x and $x + k$, with k the node spacing, the value of $\mathbf{d}$ at any location $x + i$ with $0 < i < k$ is computed as:

$$\mathbf{d}(x + i) = [\theta_1(i)\ \theta_2(i)] \begin{bmatrix} \mathbf{d}(x) \\ \mathbf{d}(x + k) \end{bmatrix}, \tag{2.1}$$

where:

$$\theta_1(i) = \frac{k - i}{k}, \tag{2.2}$$

and

$$\theta_2(i) = \frac{i}{k}. \tag{2.3}$$

Figure 2.2 shows a plot of a synthetic variable obeying this piecewise linear structure. Such a plot will always be continuous and will generally have discontinuous derivatives. We are primarily interested in the case where x, k, and i are integers; however, the above interpolation equations apply for continuous variables as well.

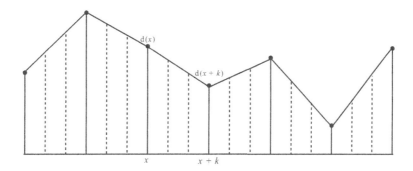

Figure 2.2: Example plot of a piecewise linear variable.

From Equation 2.1 we can construct an overall matrix equation for defining each of the M values in vector $\mathbf{d}$ from the L values in the nodal subset $\mathbf{d_L}$ where $\mathbf{d_L} = \mathbf{d}(x \in l)$ and $l = [1, 1 + k, 1 + 2k, ..., M]$. That is, we can interpolate or resample the vector $\mathbf{d_L}$ to define the higher resolution vector $\mathbf{d}$ where we have assumed that the new sampling rate is an integer multiple of the previous. Again, this assumption is for convenience and not a requirement of linear interpolation.

Defining Θ as an $M \times L$ matrix of the form:

$$\Theta = \begin{bmatrix} 1 & 0 & 0 & \cdots & 0 \\ \theta_1(1) & \theta_2(1) & 0 & \cdots & 0 \\ \vdots & \vdots & \vdots & \vdots & \vdots \\ \theta_1(k-1) & \theta_2(k-1) & 0 & \cdots & 0 \\ 0 & 1 & 0 & \cdots & 0 \\ 0 & \theta_1(1) & \theta_2(1) & \cdots & 0 \\ \vdots & \vdots & \vdots & \vdots & \vdots \\ 0 & \theta_1(k-1) & \theta_2(k-1) & \cdots & 0 \\ 0 & 0 & 1 & \cdots & 0 \\ \vdots & \ddots & \ddots & \ddots & \vdots \\ 0 & 0 & 0 & 1 & 0 \\ 0 & 0 & 0 & \theta_1(1) & \theta_2(1) \\ \vdots & \vdots & \vdots & \vdots & \vdots \\ 0 & 0 & \cdots & \theta_1(k-1) & \theta_2(k-1) \\ 0 & 0 & \cdots & 0 & 1 \end{bmatrix}, \tag{2.4}$$

we can write $\mathbf{d}$ in terms of $\mathbf{d_L}$ as:

$$\mathbf{d} = \Theta \mathbf{d_L}. \tag{2.5}$$

Program 2.1.1 provides an example of a MATLAB function that can be used to define the interpolation matrix Θ. Using the program, it is possible to construct an interpolation matrix for a fixed node spacing or for a collection of arbitrary nodes with any combination of spacings.

Program 2.1: Defining the Interpolation Matrix (Θ) for Expanding Along One Dimension

```
1   function [interp,...%left multiplying nodal data by this matrix ...
        interpolates
2            nodes] ...%positioning of the nodes
3            = defineInterpMatrix(lineLength,...%full resolution ...
                number of pixels
4            k,... %stiffness parameter describing the distance ...
                between nodes
5            nodes) %optional input prescribing node locations
6   %This function defines the interpolation matrix based on the nodes
7
8   %if the nodes are defined elsewhere, then use those
9   if(nargin==2) %otherwise define with regular spacing k
10      %such that ceil((lineLength)/k) nodes are placed at every
11      %k-pixels and always at the last index:
12      nodes = 1:k:lineLength;
13  end
14  if(nodes(end)≠lineLength);
15      nodes = [nodes,lineLength];
```

```
16  end
17  %the number of nodes and the spacing define the matrix
18  numNodes = numel(nodes);
19  interp = zeros(lineLength,numNodes);
20  %data at the nodes can be interpolated for every location by ...
        multiplying
21  %with the interpMatrix:
22  %lineDisps = interp*nodeDisps;
23
24  %using linear basis functions:
25  %the span vector describes the internodal spaces:
26  span = nodes(2:end)-nodes(1:end-1);
27  %we fill the interp matrix one 'chunk' at a time
28  %the chunk in column j will start at row nodes(j) with a value of 1
29  %and continue to subsequent rows with values dropping of 1/span ...
        each row
30  for(chunk = 1:numel(span))
31      for(row=0:(span(chunk)-1))
32          interpRow = nodes(chunk)+row;
33          firstCol = chunk;
34          secondCol = chunk+1;
35          interp(interpRow,firstCol) = 1-row/span(chunk);
36          interp(interpRow,secondCol) = row/span(chunk);
37      end
38  end
39  %The last entry is the last node and doesn't need to be interpolated
40  interp(lineLength,numNodes) =1;
41  end %%%%%%%%%%%%%%%%%%%END defineInterpMatrix
```

Building the interpolation matrix for a two-dimensional control grid is a simple Kronecker product as described in the next subsection.

2.1.2 TWO-DIMENSIONAL

In place of the vector variable $\mathbf{d}$, consider the $M \times N$ matrix $\mathbf{D}$. We define the row indices corresponding to nodes as: $l_x = 1, 1 + k, 1 + 2k, ..., M$ and the column indices correspond to nodes as: $l_y = 1, 1 + k, 1 + 2k, ..., N$ such that $\mathbf{D_L}$ contains the control points from $\mathbf{D}$:

$$\mathbf{D_L} = \mathbf{D}(x \in l_x, y \in l_y). \tag{2.6}$$

We build the higher resolution matrix $\mathbf{D}$ from $\mathbf{D_L}$ using a two-step (separable) approach to bilinear interpolation.

If the matrix $\mathbf{\Theta}_X$ is of the form shown in Equation 2.4 with M rows and the matrix $\mathbf{\Theta}_Y$ is of the same form with N rows, then we can write the full matrix $\mathbf{D}$ in terms of the control points as:

$$\mathbf{D} = \mathbf{\Theta}_X \mathbf{D_L} \mathbf{\Theta}_Y^T. \tag{2.7}$$

This expression for $\mathbf{D}$ is challenging for optimization problems involving $\mathbf{D_L}$ because of the right multiplication. To remedy this, we introduce $\bar{\mathbf{D}}_\mathbf{L}$ as a vectorized form of $\mathbf{D_L}$ (i.e., $\bar{\mathbf{D}}_\mathbf{L}$ is a column-wise concatenation of $\mathbf{D_L}$). Extending this notation, the vectorized version of the higher resolution matrix $\mathbf{D}$ (written as $\bar{\mathbf{D}}$) can be estimated using a combination matrix $\mathbf{\Theta}_{\mathbf{XY}}$ defined as the Kronecker product of the horizontal and vertical interpolating matrices such that

$$\boldsymbol{\Theta_{XY}} = \boldsymbol{\Theta}_Y \otimes \boldsymbol{\Theta}_X, \tag{2.8}$$

and

$$\bar{\mathbf{D}} = \boldsymbol{\Theta_{XY}} \bar{\mathbf{D}}_L. \tag{2.9}$$

Using Program 2.1.2, we can generate the sample interpolation matrices shown in Figure 2.3. The use of linear interpolation basis functions results in sparse, narrowly banded matrices.

Program 2.2: Example Interpolation Matrices for Expanding Along One and Two Dimensions

```
1   %display script: show sample interpolation matrices for 1D and 2D ...
        CGI
2   M = 10;%1st dimension/vector length
3   k = 3;%node spacing
4   %compute the 1D interpolation matrix
5   [interpX,nodesX] = defineInterpMatrix(M,k);
6   %confirm dimensions:
7   if(size(interpX)==[M,numel(nodesX)])
8       disp('interpolation matrix is MxL')
9       M = size(interpX,1)
10      Lx = size(interpX,2)
11  end
12  %show
13  figure
14  imshow(interpX)
15  title('1D Interpolation Matrix')
16
17  %square image example:
18  [interpXX] = kron(interpX,interpX);
19  %confirm dimensions:
20  if(size(interpXX)==[M^2,numel(nodesX)^2])
21      disp('interpolation matrix is (M^2)x(L^2)')
22      Msqrd = size(interpXX,1)
23      Lsqrd = size(interpXX,2)
24  end
25  figure
26  imshow(interpXX)
27  title('2D Interpolation Matrix for square control grid')
28
29  %rectangular image example:
30  N = 13;%2nd dimension
31  k = 3;%node spacing kept the same
32  %compute the 1D interpolation matrix
33  [interpY,nodesY] = defineInterpMatrix(N,k);
34  %confirm dimensions:
35  if(size(interpY)==[N,numel(nodesY)])
36      disp('interpolation matrix is MxL')
37      N = size(interpY,1)
38      Ly = size(interpY,2)
39  end
```

```
40
41  [interpXY] = kron(interpY,interpX);
42  %confirm dimensions:
43  if(size(interpXY)==[M*N,numel(nodesX)*numel(nodesY)])
44      disp('interpolation matrix is (M*N)x(Lx*Ly)')
45      MxN = size(interpXY,1)
46      LxLy = size(interpXY,2)
47  end
48  figure
49  imshow(interpXY)
50  title('2D Interpolation Matrix for rectangular control grid')
```

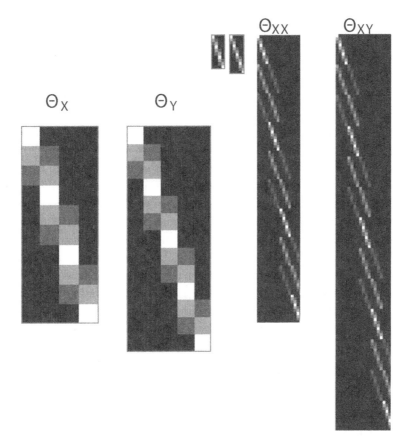

Figure 2.3: Example matrices highlighting the form of the interpolation matrices for one- and two-dimensional CGI. Program 2.1.2 can be used to generate similar sample matrices. Higher intensity indicates a higher weighting.

2.2   MULTIRESOLUTION AND ADAPTIVE CGI FORMULATIONS

Image processing problems related to motion detection frequently incorporate a multiresolution, pyramidal, or hierarchical approach to optimization. CGI can be implemented in a multiresolution framework by progressively optimizing on denser and denser grids. Factors influencing an optimization problem frequently occur at multiple resolutions. In cases where a global behavior exists, it may dominate error metrics and cost functions to the extent that more local features become lost in the noise. Minimizing a cost function at progressively finer resolutions allows the global and local behaviors (e.g., egomotion and subject motion) to be addressed.

For CGI approaches to optimization, the resolution risk is elevated in that the spacing of the control nodes can artificially emphasize one scale of problem features. Multiresolution or hierarchical CGI involves a sequential refinement of the control grid to accommodate features on smaller scales until a minimum grid size is reached. Beginning with the lowest resolution grid, the optimal solution is defined and fixed. The remaining error is then addressed with a refined grid. Generally, a quad-tree structure is utilized and the node spacing is halved each time. Figure 2.4 shows a hypothetical, three-stage refinement of a one-dimensional control grid. The final displacement variables are determined as the sum of the full resolution (interpolated) values from each of the refinement stages.

Multiresolution CGI makes it possible to address data features at multiple scales; however, the global connectivity of the control grid is retained at all resolutions and each resolution considered adds to the computational requirement. For regions with low error, where motions defined using a low resolution CGI framework minimize a given cost function well, further refinement is unnecessary, costly, and can in many cases compromise the quality of overall results. Figure 2.5 shows both hierarchical refinement of the full grid's resolution and a hypothetical, adaptively refined grid. In the adaptive refinement case, regions with high error are indicated with the bold outlines and only those portions of the grid are refined and addressed with subgrids. Alternatively, the original mesh can be structured to include more nodes in regions with higher initial error, greater structure, or saliency.

2.3   OPTIMIZATION MATHEMATICS

To this point, we have explored the structure and interpolation mathematics of CGI. In Chapter 1, the brightness constraint and the basics of optical flow were developed. In this section we detail the use of a one- and then two-dimensional CGI framework to regularize the brightness constraint with one and two degrees of freedom, respectively. To retain the framework and context for optical flow described initially, we will continue to use variables that would suggest movement (i.e., displacement over time). Figure 2.6 shows the framework we will use for the derivations. In subsequent chapters we will explore other applications.

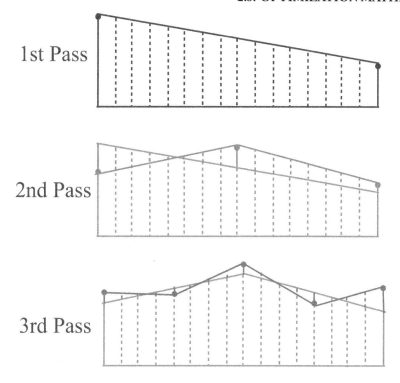

Figure 2.4: A multiresolution approach defines the final vector of variables through sequential refinement using progressively denser nodes.

2.3.1 ONE-DIMENSIONAL CONTROL GRID AND ONE DEGREE OF FREEDOM OPTICAL FLOW

The left panel of Figure 2.6 shows a theoretical displacement of a point brightness such that:

$$I(m, p) = I(m + d, p + 1). \tag{2.10}$$

Over a unit change in time, the brightness is displaced by d while retaining the same intensity. As in previous sections, m is an integer coordinate along x. The variable p is used to index integer coordinates along t.

Using the Taylor series expansion, we can generate an error equation describing deviations from the brightness constraint expressed in 2.10:

$$E(d) = \left(I(m, p + 1) - I(m, p) + d\frac{\partial I(m, p + 1)}{\partial x} \right)^2, \tag{2.11}$$

where the unknown variable d has been isolated from the image intensity term. Equation 2.11 can be directly minimized using a centered difference estimate for the partial derivative and setting the

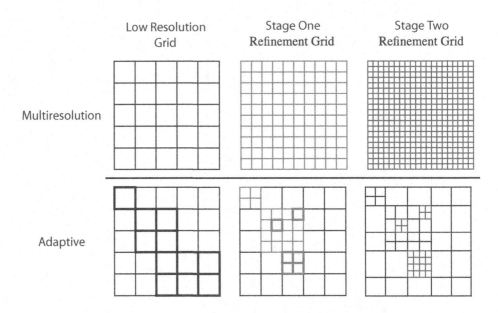

Figure 2.5:   Both the adaptive and multiresolution approaches define the final matrix of variables through sequential refinement using progressively denser nodes. Whereas the multiresolution approach refines the solutions for the full grid, only the solutions within high error regions are reevaluated in the adaptive approach.

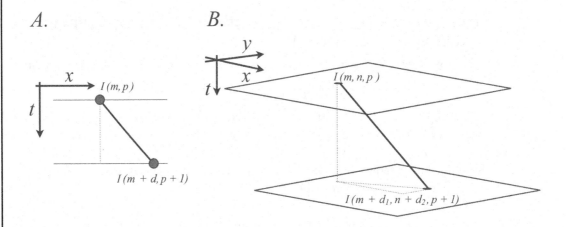

Figure 2.6:   Pixel matching framework for optical flow with (A.) one and (B.) two degrees of freedom.

error to zero such that:

$$d = 2 \left(\frac{I(m, k) - I(m, p+1)}{I(m+1, p+1) - I(m-1, p+1)} \right). \qquad (2.12)$$

This direct formulation is easily corrupted by noise when $\mathbf{I}$ is piecewise constant as is typical for many signals of interest. That is, over segments where the true signal is constant and the measured signal deviates only because of noise, the denominator will be near zero leading to spurious results.

Use of a control grid formulation to regularize Equation 2.11 effectively replaces the single equation in one unknown with an overdetermined system of equations. If we define $\bar{\mathbf{I}}(p)$ as the intensity vector with time index p and $\bar{\mathbf{I}}_{\mathbf{x}}(p)$ as the vector of all its (finite difference estimated) x partial derivates, then we can write the error (Equation 2.11) for the full vector of displacements as:

$$E(\mathbf{d}) = \left(\bar{\mathbf{I}}(p+1) - \bar{\mathbf{I}}(p) + \text{diag}[\bar{\mathbf{I}}_{\mathbf{x}}(p+1)]\mathbf{d} \right)^2. \qquad (2.13)$$

The diag[] operator places the elements of the input vector along the diagonal of an otherwise zero-valued, square matrix such that the product of the diagonal matrix and a vector is the same as the element-by-element multiplication of the vector and the input to the diag[] operator. Using a one-dimensional control grid formulation to define $\mathbf{d}$ from a subset of control nodes, $\mathbf{d_L}$ yields:

$$E(\mathbf{d_L}) = \left(\bar{\mathbf{I}}(p+1) - \bar{\mathbf{I}}(p) + \text{diag}[\bar{\mathbf{I}}_{\mathbf{x}}(p+1)]\Theta\mathbf{d_L} \right)^2. \qquad (2.14)$$

For convenience, we'll introduce a simplified notation for Equation 2.14:

$$E(\mathbf{d_L}) = \left(\bar{\mathbf{I}}_{\mathbf{t}} + \mathbf{J}\mathbf{d_L} \right)^2. \qquad (2.15)$$

The error defined by Equation 2.15 is minimized when:

$$\mathbf{J}^T(-\bar{\mathbf{I}}_{\mathbf{t}}) = \mathbf{J}^T\mathbf{J}\mathbf{d_L}. \qquad (2.16)$$

Solving for $\mathbf{d_L}$ requires inverting $\mathbf{J}^T\mathbf{J}$, a $L \times L$ matrix. This would, generally speaking, be a costly operation; however, the structure of Θ (and therefore $\mathbf{J}$) results in a tridiagonal structure for $\mathbf{J}^T\mathbf{J}$. Tridiagonal matrix inversion can be efficiently carried out using a streamlined Gaussian elimination approach and has computational demands that scale directly with the matrix size ($O(L)$). Program 2.3.1 provides an example MATLAB implementation of the tridiagonal matrix inversion or Thomas algorithm [Chapra, 1980, p. 156].

Program 2.3: Solving Systems of Equations with Tridiagonal Structure

```
1   function d = TDMA(a,b,c,e)%tridiagonal matrix solver
2   %a, b, c are the column vectors for the compressed tridiagonal ...
        matrix,
3   %e is the right vector
4   %Problem is of the form:
5   %
6   %| b(1)    c(1)     0       0       0 ... 0 |  | d(1) |    | e(1) |
7   %| a(2)    b(2)    c(2)      0       0 ... 0 |  | d(2) |    | e(2) |
8   %|  0      a(3)    b(3)    c(3)      0 ... 0 |  | d(2) |    | e(2) |
9   %|  0       0       \       \        \   0 ... 0 |x|  :   | = |  :   |
10  %|  :       :       :       :       :  | |  :   |    |  :   |
11  %|  0       0     a(n-1)  b(n-1)  c(n-1)   0 | |  :   |    |  :   |
12  %|  0       0       ...    a(n)    b(n)   c(n) | | d(n) |    | e(n) |
13  %
14  %The tridiagonal matrix algorithm performs Gaussian elimination ...
        using a
15  %standard sequence capitalizing on the structure of the ...
        coefficient matrix
16  n = length(b); % n is the number of rows
17
18
19  %Stage 1: Modify the problem to be of the form:
20  %
21  %| 1      c'(1)     0       0       0 ... 0 |  | d(1) |    | e'(1) |
22  %| 0       1      c'(2)     0       0 ... 0 |  | d(2) |    | e'(2) |
23  %| 0       0       1      c'(3)     0 ... 0 |  | d(2) |    | e'(2) |
24  %| 0       0       \       \        \   0 ... 0 |x|  :   | = |  :   |
25  %| :       :       :       :       :       : | |  :   |    |  :   |
26  %| 0       0       0       1      c'(n-1)  : | |  :   |    |  :   |
27  %| 0       0       ...     0       0       1 | | d(n) |    | e'(n) |
28  %
29  % updating c and e to c' and e' in place:
30  if(b(1))% Division by zero risk
31      c(1) = c(1) / b(1);
32      e(1) = e(1) / b(1);
33  else %pseudo invert (ie. replace the infs and NaNs with 0)
34      c(1) = 0;
35      e(1) = 0;
36  end
37
38  for i = 2:n
39      denom = b(i) - (c(i-1)*a(i));
40      if(denom) %nonzero case
41          coeff = 1 / denom;
42      else %replacing with 0 if undefined
43          coeff = 0;
44      end
45      c(i) = c(i)*coeff;
46      e(i) = (e(i) - e(i-1) * a(i)) * coeff;
47  end
48
49  %Stage 2:Back substitute to solve for each value in d
50  d(n) = e(n);
51  for i = n-1:-1:1
52      d(i) = e(i) - c(i) * d(i + 1);
```

```
53   end
54   end
```

2.3.2   TWO DIMENSIONAL CONTROL GRID AND ONE DEGREE OF FREEDOM OPTICAL FLOW

We now consider a special case of the right pane of Figure 2.6 where the displacement along x, d_1, is unknown and the displacement along y is fixed at $d_2 = 0$. This would be, for example, a reasonable assertion in the case of a video camera with the line of sight perpendicular to car motion along a highway. In such a scenario, Equation 2.11 remains relatively untouched and the foundational problem remains uniquely determined:

$$E(d_1) = \left(I(m, n, p + 1) - I(m, n, p) + d_1 \frac{\partial I(m, n, p + 1)}{\partial x} \right)^2. \qquad (2.17)$$

As in the one-dimensional CGI approach (formalized in Equation 2.14) we can extend Equation 2.17 to describe the error for the full matrix of displacements using a vectorized version of the input signals and the two-dimensional interpolation matrix defined in Equation 2.8:

$$E(\mathbf{d_{1L}}) = \left(\bar{\mathbf{I}}(p + 1) - \bar{\mathbf{I}}(p) + \text{diag}[\bar{\mathbf{I}}_\mathbf{x}(p + 1)]\mathbf{\Theta_{XY}d_{1L}} \right)^2. \qquad (2.18)$$

As before, we define $\bar{\mathbf{I}}(p)$ as the signal vector with time index p; however, in the two-dimensional case the vector is a concatenated version of the signal matrix. The vector $\bar{\mathbf{I}}_\mathbf{x}(p)$ remains the vector containing the finite-difference estimated x partial derivates; it is important to note that the vectorization occurs after the derivative calculation.

   Given Equation 2.18, the only structural change to the problem is in the form of the interpolating matrix. Reviewing Figure 2.3, the two-dimensional interpolation matrix remains sparse but with a wider data band. As a result, the coefficient matrix to be inverted (now an $L_x L_y \times L_x L_y$ matrix) also exhibits a wider banding and the Thomas algorithm presented in Program 2.3.1 is no longer applicable. If we retain the shorthand expression for the least-squares minimization problem shown in Equation 2.16, where $\mathbf{J}$ includes the interpolation matrix $\mathbf{\Theta_{XY}}$, then the resulting coefficient matrix $\mathbf{J}^T\mathbf{J}$ will be a symmetric, block tridiagonal matrix. Figure 2.7 shows this graphically using the sample $\mathbf{\Theta_{XY}}$ matrix from Figure 2.3. The block tridiagonal structure invites efficient analytical solution algorithms [Jain et al., 2007]. Alternatively, the local dominance (matrix values decrease away from the central diagonal) can be capitalized on using an iterative, sliding window technique [Frakes et al., 2001].

   The framework described is limited in that the second degree of freedom generally incorporated in two-dimensional motion estimation has been neglected. As such, the proposed framework does not address the aperture problem of optical flow. In the next subsection, we extend our scope to include the two degrees of freedom represented in the traditional brightness constraint error model.

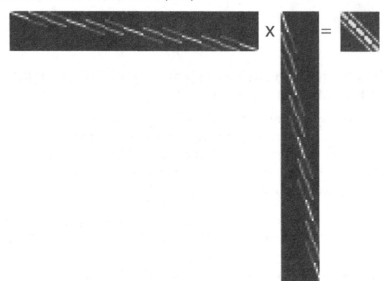

Figure 2.7: Example matrices highlighting the form of the matrix to be inverted given a two-dimensional control grid. Brighter pixels indicate higher weighting.

2.3.3   TWO-DIMENSIONAL CONTROL GRID AND TWO DEGREES OF FREEDOM OPTICAL FLOW

In the general case, the right pane of Figure 2.6 depicts the scenario where the displacements along both x and y (d_1 and d_2, respectively) are unknown. In such a scenario, Equation 2.11 takes on the form of traditional optical flow:

$$E(d_1, d_2) = \left(I(m, n, p + 1) - I(m, n, p) + d_1 \frac{\partial I(m, n, p + 1)}{\partial x} + d_2 \frac{\partial I(m, n, p + 1)}{\partial y} \right)^2.$$

$$(2.19)$$

As in previous approaches, we can extend Equation 2.19 to describe the error for the full matrix of displacements using a vectorized version of the input signals and the two-dimensional interpolation matrix defined in Equation 2.1:

$$E(\mathbf{d_{1L}}, \mathbf{d_{2L}}) = \left(\bar{\mathbf{I}}(p + 1) - \bar{\mathbf{I}}(p) + \text{diag}[\bar{\mathbf{I}}_x(p + 1)]\mathbf{\Theta_{XY}}\mathbf{d_{1L}} + \text{diag}[\bar{\mathbf{I}}_y(p + 1)]\mathbf{\Theta_{XY}}\mathbf{d_{2L}} \right)^2.$$

$$(2.20)$$

To reduce Equation 2.20 to a three variable form similar to Equation 2.15, we propose defining $\mathbf{J}$ and $\mathbf{d_L}$ such that:

$$\mathbf{J}\mathbf{d_L} = \text{diag}[\bar{\mathbf{I}}_x(p + 1)]\mathbf{\Theta_{XY}}\mathbf{d_{1L}} + \text{diag}[\bar{\mathbf{I}}_y(p + 1)]\mathbf{\Theta_{XY}}\mathbf{d_{2L}}.$$

$$(2.21)$$

In order to keep the structure of $\mathbf{J}^T\mathbf{J}$ as narrow as possible, we interleave the horizontal and vertical displacement contributions such that:

$$
\begin{aligned}
\mathbf{d_L} = [&\mathbf{d_{1L}}(1), \mathbf{d_{2L}}(1), \mathbf{d_{1L}}(2), \mathbf{d_{2L}}(2), ..., \\
&\mathbf{d_{1L}}(L_x), \mathbf{d_{2L}}(L_x), \mathbf{d_{1L}}(L_x+1), \mathbf{d_{2L}}(L_x+1), ..., \\
&\mathbf{d_{1L}}(L_xL_y), \mathbf{d_{2L}}(L_xL_y)]^T.
\end{aligned}
\tag{2.22}
$$

Similarly, the columns of $\mathbf{J}$ alternate between columns from the product $\mathrm{diag}[\bar{\mathbf{I}}_\mathbf{x}(p+1)]\Theta_{\mathbf{XY}}$ and the product $\mathrm{diag}[\bar{\mathbf{I}}_\mathbf{y}(p+1)]\Theta_{\mathbf{XY}}$. With this approach, $\mathbf{J}^T\mathbf{J}$ will retain the symmetric, block tridiagonal structure seen in the two-dimensional, one degree of freedom case. As such, analytical solutions are tractable; however, as with the one degree of freedom case, sliding window and other iterative approaches may be applied. In many cases (some of which will be covered in more depth in subsequent chapters), maintaining the global connectivity of the control grid does not necessarily result in an ideal solution. Numerical approaches that emphasize the local optimality of the solution and allow for deviations from a pure CGI framework can allow additional flexibility and improve the fidelity of the estimated motion field. Iterative frameworks may also lend themselves more readily to local refinement of the control grid and adaptive meshes.

2.4   SYMMETRIC IMPLEMENTATIONS

Prior to addressing specific applications of the CGI approach to optical flow and other similar problems, it is useful to extend our commentary on the computation of the partial derivative terms, $\mathbf{I_x}$, $\mathbf{I_y}$, and $\mathbf{I_t}$. Thus far we have provided a centered finite difference as the method for computing $\mathbf{I_x}$ and a backward difference for computing $\mathbf{I_t}$. Depending on which data are available and what assertions are being made regarding the duration of a given trajectory, alternative approaches to estimating the derivatives may be more appropriate. Furthermore, specific assumptions and formulations can result in a framework with useful symmetric properties.

Generally, speaking, optical flow algorithms do not produce symmetric motion fields. That is, the motion field estimated from time frame t to time frame $t+1$ is not equivalent to the inverted motion field estimated from time $t+1$ to time t. Figure 2.4 shows a centered framework that enforces a symmetric solution. Rather than a directed solution from a source image to a target image, the centered framework describes motion through an intermediate frame. The brightness constraint is applied to all three frames such that:

$$
\mathbf{I}(m-d_1, n-d_2, p-\Delta_t) = \mathbf{I}(m, n, p) = \mathbf{I}(m+d_1, n+d_2, p+\Delta_t).
\tag{2.23}
$$

Taylor series expansion of Equation 2.23 yields (left hand side):

$$
\mathbf{I}(m-d_1, n-d_2, p-\Delta_t) = \mathbf{I}(m, n, p) - d_1\mathbf{I_x}(m, n, p) - d_2\mathbf{I_y}(m, n, p) - \Delta_t\mathbf{I_t}(m, n, p),
\tag{2.24}
$$

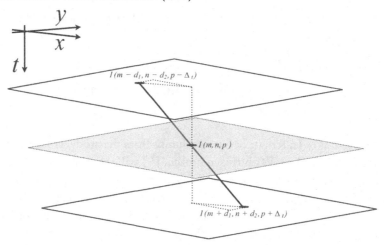

Figure 2.8: Framework for symmetric optical flow implementations in two-dimensions with two degrees of freedom.

and (right hand side):

$$\mathbf{I}(m + d_1, n + d_2, p + \Delta_t) = \mathbf{I}(m, n, p) + d_1\mathbf{I_x}(m, n, p) + d_2\mathbf{I_y}(m, n, p) + \Delta_t\mathbf{I_t}(m, n, p), \tag{2.25}$$

where $\mathbf{I_x}$, $\mathbf{I_y}$, and $\mathbf{I_t}$ are discrete approximations to the partial derivatives. The spatial indexing coordinates of the displacement terms are not included for space, and all displacements are tied to the central grid (e.g., $d_1 = \mathbf{d_1}(m, n, p)$ in the example shown). The corresponding error equation (using the squared difference) is:

$$E(d_1, d_2) = \left[\frac{d_1}{\Delta_t}\mathbf{I_x}(m, n, p) + \frac{d_2}{\Delta_t}\mathbf{I_y}(m, n, p) + \mathbf{I_t}(m, n, p) \right]^2. \tag{2.26}$$

CGI can be used to minimize global error in a connected framework using the matrix equation for the overall error associated with the displacement field $[\mathbf{d_1}, \mathbf{d_2}]$. In contrast to previously described approaches to motion estimation, the frame in which and method by which the partial derivatives are computed is the essential choice in implementing the symmetric approach. As highlighted in subsequent chapters, we are frequently interested in the scenario wherein the intermediate frame is yet to be defined, necessitating that its derivatives be estimated using the surrounding, measured data.

2.5   SUMMARY

CGI refers to a relatively generic framework for implementing spatial transformations and can be used to transform optimization problems into a framework with fewer unknown variables. In this chapter, we explored use of the CGI framework to constrain problems related to the brightness constraint and optical flow. The remainder of this book will detail several applications of CGI in the context of registration as well as registration-based interpolation.

CHAPTER 3

Application of CGI to Registration Problems

The objective of optical flow and any application of the brightness constraint is identifying a one-to-one match between two datasets based on intensity. Use of a control grid to regularize the cost function for optical flow can provide robustness to noise in uniquely determined scenarios and constrain underdetermined optimization problems. While the origins of optical flow (and its most common applications) are in the context of video and motion estimation, the fundamental mathematics of intensity matching have widespread applications. In this chapter we explore several contexts for intensity-based registration.

3.1   REGISTRATION OF ONE-DIMENSIONAL DATA: INTER-VECTOR REGISTRATION

In Section 2.3.1 we described use of a one-dimensional control grid formulation for determining a single displacement or offset variable that optimally matches two vectors of data. The matching equation used in that framework (Equation 2.10 reproduced and renumbered here) is:

$$\mathbf{I}(m, p) = \mathbf{I}(m + d, p + 1), \tag{3.1}$$

where $\mathbf{I}$ is a measure of brightness, m is a spatial indexing variable, and p is a temporal indexing variable. In this section we will outline other applications for one-dimensional control grids with similar optimization mathematics.

3.1.1   DYNAMIC TIME WARPING

Dynamic time warping describes a specific approach to registering two time-dependent sequences. Given specific assumptions or restriction (e.g., the order of events is the same in both sequences), the alignment that minimizes the difference between the two sequences is determined. Originally applied to speech patterns [Itakura, 1975], dynamic time warping or DTW has been applied in a wide range of applications [Kovacs-Vajna, 2000, Legrand et al., 2008, Rath and Manmatha, 2003]. The classical approach to DTW is perhaps best explained as it is commonly visualized. Referring to Figure 3.1, a point-by-point comparison of the two signals is computed (for example by measuring the absolute or square difference in the signals) at all possible pairings. The 'optimal' path through the matrix of all pair differences is then identified. Starting from time zero in both signals (the assumed optimal first pair) and navigating to the final time point in both signals (the presumed final pair), the best path is selected to minimize the total difference between the two signals (normalized for path length). Other restrictions or assumptions (e.g., a preference for the shortest path) are also incorporated to make the search space more tractable. Alternatively, much like block matching and optical flow, parametric approaches to identifying the optimal path have also been utilized [Eilers, 2004, Keogh and Pazzani, 2001]. The one-dimensional framework for optical flow based on the brightness constraint in Equation 3.1, and its Taylor series expansion, can be used directly for DTW applications.

The broad applicability of the DTW approach demonstrates that the signal and constraint need not be brightness. For example, the DTW application demonstrated in Figure 3.1 is the alignment of the voltage signals from two heartbeats in a synthetic electrocardiogram (ECG) waveform. Constraining the amplitude of the ECG can be accomplished with the same optimization mathematics outlined in the previous chapter and demonstrated in Program 3.1.1 below. The implementation provided iterates over multiple resolutions and a sample output is shown in Figure 3.2.

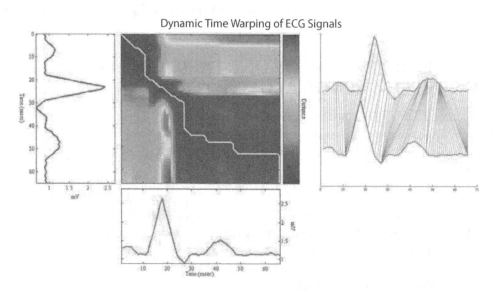

Figure 3.1: Sample distance matrix and optimal path for dynamic time warping between two ECG signals. Also shown are the points of correspondence defined by the 'optimal' path.

Program 3.1: Sample DTW Algorithm based on CGI and Optical Flow

```
1  function [displacements] ...%the computed displacements
2                           = linePair_timeWarp(linePair,... %vectors ...
                             to match
3                                        k) %node spacing(s)
4
```

```
5  %Computes the displacments that map one time signal to another ...
       based on
6  %matched signal strength
7
8  %linePair is the array [line1;line2]
9  %k defines the node spacing and can be a vector specifying a ...
       series of
10 %spacings (should monotonically decrease)
11
12
13 %from the input define:
14 M = size(linePair,2);%line length
15 pyramidHeight = length(k);%the number of steps in the pyramid
16
17 %initialize:
18 displacements = zeros(M,1);
19 source  = linePair(1,:);
20 matches = linePair(2,:);
21
22 %prep the figure to plot the matches:
23 dispMatches = figure('Name','Point Correspondence','Color',[1 1 1]);
24
25 for(step = 1:pyramidHeight) %progress through resolution pyramid
26 %define the nodes and intepolation matrix
27 [interp,nodes] = defineInterpMatrix(M,k(step));
28 L = length(nodes);
29 %force the first and last displacements to be zero:
30 interp(:,1) = 0;
31 interp(:,end) = 0;
32
33 %determine the linePair derivatives
34 I_t = [matches(:)-source(:)];
35 I_x = [matches(2)-matches(1),...
36        0.5*[matches(3:end)-matches(1:end-2)],...
37        matches(end)-matches(end-1)]';
38
39 %structure the matrices
40 J = diag(I_x)*interp;
41 JtransJ = J'*J;
42 RHS = J'*(-1*I_t);
43
44 %solve for the displacments at the grid points:
45 %split the matrix into the diagonal components:
46 a = [0 JtransJ(2:L+1:end)];
47 b = JtransJ(1:L+1:end);
48 c = [JtransJ(L+1:L+1:end) 0];
49 %note that e = RHS;
50 nodalDisps = TDMA(a,b,c,RHS);
51
52 %interpolate to find the full solution and add to existing ...
       displacmenets
53 displacements = displacements + interp*nodalDisps';
54
55 %convert the computed displacements to matched locations, enforce ...
       'rules'
56 %force the range to stay inside measured time:
```

```matlab
57    targetLocations = min(max(double(1:M)+displacements',1),M);
58    %replace any overlaps with singularities:
59    prevLoc = targetLocations(1);
60    for ndx=2:M
61        if(targetLocations(ndx)<prevLoc)
62            targetLocations(ndx) = prevLoc;
63        end
64        prevLoc = targetLocations(ndx);
65    end
66    displacements = [targetLocations-double(1:M)]'; %update the ...
          displacements
67
68    %normalize the signals for plotting:
69    line1 = linePair(1,:);
70    normed1 = (line1-mean(line1))./std(line1);
71    line2 = linePair(2,:);
72    normed2 = (line2-mean(line2))./std(line2);
73    %define the matches in the target line:
74    matches = interp1(1:M,normed2,targetLocations,'pchip',0);
75
76    %%%% PLOTTING UTILITIES %%%%
77    %hard-coded, defines amount of space between the two signals in plot
78    separate = 5;
79    %plot the normalized signals
80    subplot(pyramidHeight,1,step)
81    hold on;
82    title(['Multiresolution Pass ',num2str(step),', k = ...
          ',num2str(k(step))],...
83            'FontSize',14,'FontName','Times');
84    plot(normed1,'r-','LineWidth',2);
85    plot(normed2-separate,'b-','LineWidth',2);
86    if(step≠pyramidHeight)
87      set(gca, 'ytick',[],'xtick',[],'FontName','Times','FontSize',14);
88    else
89      set(gca, 'ytick',[],'FontName','Times','FontSize',14);
90    end
91    %connect the matches
92    for ndx=1:M
93        origin = ndx;
94        destination = targetLocations(ndx);
95        h=line([origin destination], ([normed1(ndx) ...
              matches(ndx)-separate]));
96        set(h,'LineWidth',1,'color',[0.5 0.5 0.5]);
97    end
98    %%%%%%%%
99
100   end
```

DTW can be used to align any pair of time-dependent signals. In preparation for the next subsection, we introduce one more specific example from the field of analytical chemistry. Most people have conducted a chromatography experiment as a child. A chemical compound or mixture is drawn through a medium that allows the different components to travel at different rates thus separating them in space over time. A common example is separating the different colors in ink or dye. Figure 3.3 shows a hypothetical example with the solvent fronts marked. For analytical chemists, the exact travel times of each component are essential; however, the bands in the resulting

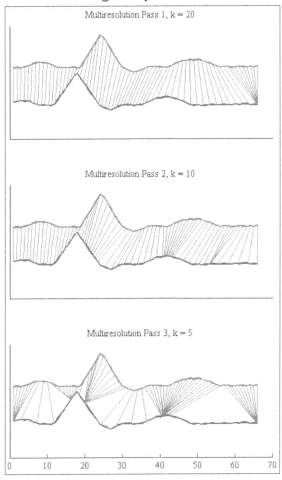

Figure 3.2: Results from a multiresolution approach to dynamic time warping based on CGI optical flow.

chromatogram are subject to distortions and are not uniform or straight. If each column of pixels in the chromatogram is considered as a one-dimensional signal (intensity versus time), DTW can be utilized to align the bands. In this application, DTW registers each line of signals based on matching intensities and obeying basic constraints. The registered signals are aligned in time and the expectation for alignment is based on physics (i.e., it is reasonable to treat each line of pixels as an independent repetition of the same experiment; a different measurement of the same process). In the next section, we will explore other registration problems related to matching intensities across a pair of one-dimensional signals, lifting the restriction that the signals be time dependent or repeated measures of the same physical phenomenon.

Figure 3.3: A synthetic chromatogram with example solvent fronts marked.

3.1.2 ISOPHOTE IDENTIFICATION

Functionally, DTW in the case of chromatogram alignment amounts to identifying isophotes (curves of constant intensity) within a two-dimensional image. This is clear in looking at the hypothetical solvent front markings shown in Figure 3.3. In this subsection, we look to use a similar, one-dimensional CGI approach to optical flow with one degree of freedom to identify isophotes within a generic digital image. We continue to allow a single degree of freedom; however, the two-dimensional image matrix is addressed both as a set of row vectors and a set of column vectors. Here the isophote is approximated locally with a line of constant intensity such that:

$$\mathbf{I}(m, n) = \mathbf{I}(m + t\alpha, n + t), \qquad -1 < t < 1 \tag{3.2}$$

and

$$\mathbf{I}(m, n) = \mathbf{I}(m + u, n + u\beta), \qquad -1 < u < 1. \tag{3.3}$$

This implies that the isophote can be locally described by its tangent.

For our control grid formulation, we will look to identify a pair of displacements d_1 and d_2 at every pixel that registers the pixel of interest to the best match in the neighboring columns and rows. Figure 3.4 shows this in the context of the chromatogram example. Rewriting Equations 3.2 and 3.3 with normalized displacements yields:

$$\mathbf{I}(m, n) = \mathbf{I}(m \pm d_1, n \pm 1) \tag{3.4}$$

and

$$\mathbf{I}(m, n) = \mathbf{I}(m \pm 1, n \pm d_2). \tag{3.5}$$

Identifying the values for d_1 and d_2 that best satisfy these equalities constitutes two separate optimizations problems. Using the problem classifications introduced in Chapter 2, each problem is symmetric and has one degree of freedom. As a result, each problem is inherently overdetermined with a single unknown and two equations; however, the initial formulation is not tractable. If d_1 and d_2 are allowed to be any real valued number (or even real valued over some magnitude range), an exhaustive search to find the closest match is not possible. As with block matching, a search-based

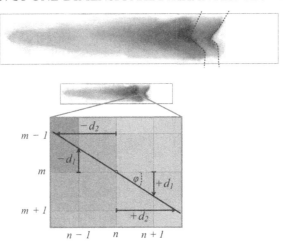

Figure 3.4: The isophote description framework is shown in the context of the chromatogram example.

procedure could be employed if the set of possible displacements is quantized. Alternatively, if the intensity function is assumed to be locally planar, we can use a first-order Taylor series approximation to expand the constraint equations to:

$$\mathbf{I}(m, n) = \mathbf{I}(m, n) \pm d_1 \frac{\partial \mathbf{I}(m, n)}{\partial x} \pm \frac{\partial \mathbf{I}(m, n)}{\partial y} \tag{3.6}$$

and

$$\mathbf{I}(m, n) = \mathbf{I}(m, n) \pm \frac{\partial \mathbf{I}(m, n)}{\partial x} \pm d_2 \frac{\partial \mathbf{I}(m, n)}{\partial y}. \tag{3.7}$$

To find d_1 and d_2 that best satisfy Equations 3.6 and 3.7 we formulate the error expressions:

$$E(d_1, m, n) = \left[d_1 \mathbf{I_x}(m, n) + \mathbf{I_y}(m, n) \right]^2 \tag{3.8}$$

and

$$E(d_2, m, n) = \left[d_2 \mathbf{I_y}(m, n) + \mathbf{I_x}(m, n) \right]^2, \tag{3.9}$$

where $\mathbf{I_x}$ and $\mathbf{I_y}$ are discrete approximations of the x and y partial derivatives, respectively. Assuming that we seek to effectively identify isophotes regardless of their orientation, selection of a maximally rotation invariant derivative kernel, for example, the Scharr modified Sobel kernel [Bradski and Kaehler, 2008, p. 150], is indicated. Program 3.1.2 computes the Scharr derivatives (using the coordinate convention outlined in Subsection 1.4) and generates an output similar to that shown in Figure 3.5.

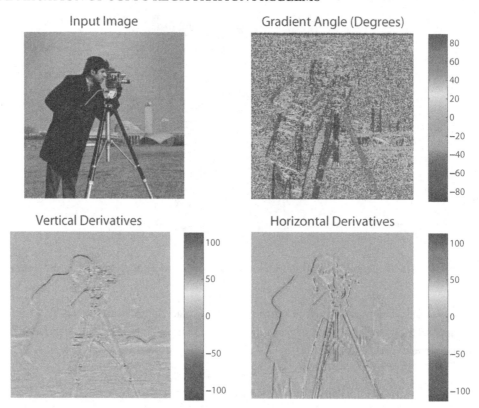

Figure 3.5: The *x* and *y* partial derivatives computed with the Scharr derivative kernels are shown along with the gradient angle.

Program 3.2: Computation of the derivatives with the Scharr (modified Sobel) kernel

```
1  function [Ix, ...%derivative along x/first/column/vertical dimension
2            Iy, ...%derivative along y/second/row/horizontal dimension
3            phi]...%the counter-clockwise gradient angle
4                  = scharrDerivatives(testImage,...%input 2D image ...
                      matrix
5                                  clean,... %eliminate small ...
                                    derivatives
6                                  visualize)   %show the images
7
8  %computes the partial x and y derivatives using the Scharr ...
      derivatives
9  %optional clean arguement removes small derivatives,
10 %   if clean = 0 or empty then values are untouched, otherwise ...
      derivatives
11 %   with magnitudes less than clean*max(testImage) are set to zero.
12 %visualize generates a plot of the image, derivatives, and ...
      gradient angle
```

```
13   %no inputs runs as a demo using MATLAB packaged cameraman image
14
15   %Defaults
16   if(nargin<3)
17       visualize = 0; %default is not to show figures
18       if(nargin<2)
19           clean = 0; %default is to not clean
20           if(nargin<1)
21               testImage = imread('cameraman.tif');%run as demo
22               visualize = 1; %show figures for demo
23           end
24       end
25   end
26
27   %convert the input to double if necessary
28   if(isinteger(testImage))
29       testImage = double(testImage);
30   end
31
32   %Begin with simple subtraction along the individual rows and columns
33   %use centered difference where possible and forward differences ...
         when not
34
35   %differences going along columns:
36   Icol = [(testImage(2,:)-testImage(1,:)); ...
37           (1/2)*(testImage(3:end,:)-testImage(1:end-2,:)); ...
38           (testImage(end,:)-testImage(end-1,:))];
39   %
40   Irow = [(testImage(:,2)-testImage(:,1)) ...
41           (1/2)*(testImage(:,3:end)-testImage(:,1:end-2)) ...
42           (testImage(:,end)-testImage(:,end-1))];
43
44
45   %extend the derivatives to be from 3x3 Scharr kernels
46   Ix = 1/16*[3*Icol(:,1:end-2)+10*Icol(:,2:end-1)+3*Icol(:,3:end)];
47   Ix = [Icol(:,1), Ix, (Icol(:,end))];
48
49   Iy = (1/16)*[3*Irow(1:end-2,:)+10*Irow(2:end-1,:)+3*Irow(3:end,:)];
50   Iy = [Irow(1,:); Iy; Irow(end,:)];
51
52   %remove small derivatives:
53   if(clean)
54       Ix(abs(Ix)<clean*max(testImage(:)))=0;
55       Iy(abs(Iy)<clean*max(testImage(:)))=0;
56   end
57
58   %compute the pixel-by-pixel gradient angle phi
59   phi = atand(-1*(Iy)./(Ix));
60   %note: NaNs are likely to occur.
61
62   %sample plots
63   if(visualize)
64       figure('Name','Derivative Images','Color',[1 1 1]);
65       subplot('position',[0.03 0.5 0.4 0.4]);
66       imshow(repmat(testImage,[1,1,3])/max(testImage(:)),[])
67       title('Input Image','FontSize',14,'FontName','Times')
```

```
68          subplot('position',[0.5 0.5 0.4 0.4]);
69          imshow(phi,[-90,90])
70          colorbar('FontSize',14,'FontName','Times')
71          title('Gradient Angle ...
                (Degrees)','FontSize',14,'FontName','Times')
72          subplot('position',[0.05 0.02 0.4 0.4]);
73          imshow(Ix,[min([Ix(:);Iy(:)]),max([Ix(:);Iy(:)])])
74          colorbar('FontSize',14,'FontName','Times')
75          title('Vertical Derivatives','FontSize',14,'FontName','Times')
76          subplot('position',[0.5 0.02 0.4 0.4]);
77          imshow(Iy,[min([Ix(:);Iy(:)]),max([Ix(:);Iy(:)])])
78          colorbar('FontSize',14,'FontName','Times')
79          title('Horizontal ...
                Derivatives','FontSize',14,'FontName','Times')
80          colormap('jet')
81      end
```

Also shown in Figure 3.5 is the gradient angle image. This is the direct approximation to ϕ (as defined in Figure 3.4) from either Equation 3.8 or Equation 3.9. Specifically, forcing the error defined in Equation 3.8 to zero results in:

$$\mathbf{D}_1(m, n) = \frac{-\mathbf{I}_y(m, n)}{\mathbf{I}_x(m, n)}. \tag{3.10}$$

Similarly, forcing the error in Equation 3.9 to zero yields:

$$\mathbf{D}_2(m, n) = \frac{-\mathbf{I}_x(m, n)}{\mathbf{I}_y(m, n)}. \tag{3.11}$$

Using either displacement to define the angle ϕ gives:

$$\tan(\boldsymbol{\phi}(m, n)) = \mathbf{D}_1(m, n), \tag{3.12}$$

or equivalently:

$$\cot(\boldsymbol{\phi}(m, n)) = \mathbf{D}_2(m, n). \tag{3.13}$$

The angle ϕ is computed as:

$$\boldsymbol{\phi} = \tan^{-1}\left(\frac{-\mathbf{I}_y(m, n)}{\mathbf{I}_x(m, n)}\right), \tag{3.14}$$

and plotted in degrees in Figure 3.5. In natural images, the intensity derivatives are frequently near-zero (images are piecewise stationary). Directly computing ratios of the gradients (or ϕ) results in noisy results. These can be improved if small derivatives are set to zero. In Figure 3.6, derivatives with absolute values less than 1% if the peak image intensity were forced to zero before computing the gradient angle ϕ. This cleaning procedure is essential when using ϕ to track lines of constant intensity in the image as noisy values will contribute severe distortions.

Program 3.1.2 provides a simple MATLAB approach to tracking image isophotes using a displacement term (e.g., $\mathbf{d}_2$ or $\cot(\boldsymbol{\phi})$). In contrast to the displacements, which provide a local, linear

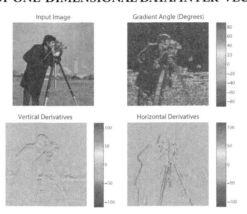

Figure 3.6: The *x* and *y* partial derivatives computed with the Scharr derivative kernels are clipped to remove small magnitude ($< 1\%$ maximum signal intensity) results prior to computing the gradient angle. The cleaned areas are shown in dark navy.

approximation of the isophote crossing through each pixel location, Program 3.1.2 redefines the displacements at the matched locations progressively constructing continuous isophotes from the bottom of the image up (top-down and left-right or right-left are equally possible). The approach is similar to that proposed by Wang et al. [2002]; however, in that work, the tracking is done from image intensities directly whereas we use the displacements. Figure 3.7 shows the tracked isophotes for the cameraman image using the raw and cleaned derivatives (shown in Figures 3.5 and 3.6, respectively) to define the displacement term. Clearly, the isophote tracking approach fails with the raw derivatives; however, in the cleaned approach, only isophotes near strong edges are retained. Use of a control grid framework allows meaningful isophotes to be identified through the noise.

Program 3.3: Using the displacements, plot an isophote sketch of the image

```
1  function isophotesFromDisplacments(d,... %displacements
2                                  k) %space between displacments
3
4  %Use the displacements to plot a line drawing of an image
5  %assumes displacments are horizontal offsets/matches made from ...
      row-to-row
6  %k indicates spacing between displacements, k=1 indicates same as ...
      line
7  %spacing.
8  if(nargin<2)
9      k=1;%default
10 end
11
12 [M,N] = size(d);
13 %protection against NaNs and Infs (e.g. when d = cotd(phi) and ...
      phi=0).
```

A.

B.

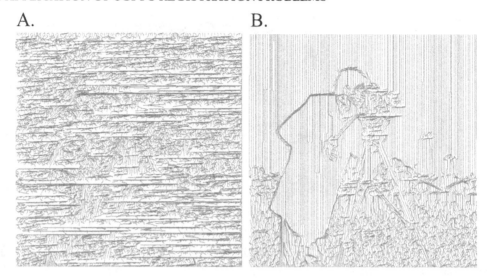

Figure 3.7: Both image panes show approximate isophote sketches of the cameraman image. (A.) Results from raw image derivatives. (B.) Results from image derivatives where derivatives with magnitudes less than 1% of the maximum image intensity have been set to zero and associated displacements have been set to zero as well.

```
14   d(isnan(d))=0;
15   d(isinf(d))=0;
16
17   %%%% PLOTTING UTILITIES %%%%
18   isophotePlot = figure('Name','Isophote Tracking','Color',[1 1 1]);
19   title(['Isophotes'],'FontSize',14,'FontName','Times');
20   hold on;
21   %connect the matches
22   %start at evenly spaced points along the bottom:
23   lattice = 1:k:N*k;
24   origin = lattice;
25   for m=M:-1:1  %step through the rows starting at the bottom
26       %interpolate to find the displacements where the isophotes ...
             intersect
27       %the current line
28       dispAtOrigin = interp1(lattice,d(m,:),origin,'linear',0);
29       %the displacements tell us where in the previous line the ...
             isophote
30       %intersects
31       destination = origin - dispAtOrigin;
32       %check for bounds on the full set:
33       destination = min(max(destination,lattice(1)),lattice(end));
34
35       %Begin plotting the isophotes connecting the rows m and m-1
36       %up from the bottom:
37
```

```
38      %initialize:
39      prevLoc = destination(1);
40      newOrigin = origin;
41      ndx = 2;
42
43      %plot each isophote separately:
44      for n=2:N
45        if(destination(n)<prevLoc) %check if paths have merged
46          destination(n) = prevLoc; %keep just one in next set
47        elseif(destination(n)>(prevLoc + 3*k)) %check for 'large' gaps
48          newOrigin(ndx) =  0.5*(prevLoc + destination(n));% seed a ...
            new point
49          ndx = ndx+1;
50          newOrigin(ndx) = destination(n); %retain the existing ...
            point as well
51          ndx = ndx+1;
52        else
53          newOrigin(ndx) = destination(n); %no modifications
54          ndx = ndx+1;
55        end
56        prevLoc = destination(n);
57        %add to graph:
58        h=line([origin(n) destination(n)], [M-m-1 M-m]);
59        set(h,'LineWidth',1,'color',[0.5 0.5 0.5]);
60      end
61      %adjust for next row
62      N = ndx;
63      origin = [newOrigin(1:N-1) lattice(end)];
64   end
65   %%%%%%%%
66   set(gca, 'ytick',[],'xtick',[],'FontName','Times','FontSize',14);
67   end
```

Noting that Program 3.1.2 uses the d_2 displacement variable to plot the isophotes, we return to the associated cost function, Equation 3.9, and use a collection of control points to reduce the number of variables being optimized for:

$$E(\mathbf{d}_{2L}(m)) = \left(\bar{\mathbf{I}}_x(m) + \text{diag}[\bar{\mathbf{I}}_y(m)]\Theta\mathbf{d}_{2L}(\mathbf{m}) \right)^2. \tag{3.15}$$

The vectors $\bar{\mathbf{I}}_x(m)$ and $\bar{\mathbf{I}}_y(m)$ contain all of the x and y partial derivatives for row m and Θ is the interpolation matrix such that the displacements at the control nodes $(\mathbf{d}_{2L}(\mathbf{m}))$ are related to the full set as $\mathbf{d}_2(\mathbf{m}) = \Theta\mathbf{d}_{2L}(\mathbf{m})$. It is important to note that in this symmetric implementation the displacements are assigned to the same pixel that the derivatives are defined at. Additionally, while the optimization for the d_2 displacements accesses rows of data at a time, optimization for the d_1 displacements uses columns (vector variables are always treated as having a vertical orientation). Using Programs 2.1.1, 2.3.1, and 3.1.2 as subfunctions, we provide Program 3.1.2 to generate a displacement pair for each pixel in an input image. The displacements computed using Program 3.1.2 and both raw and cleaned derivatives are used to generate Figure 3.8. As with the directly computed displacements, the removal of small magnitude derivatives eliminates some spurious isophotes as well as some meaningful ones. In contrast to the directly computed results, the CGI results using the raw derivatives show many isophotes with significant path lengths that track well with the visible features of the cameraman image. Additionally, in the areas of the man's elbow and lower jacket, detailed contour information difficult to see in the original image is apparent. In both sketches based on the CGI generated displacements, isophotes are tracked and remain further inside structures where image intensities are smoother, as opposed to being pulled exclusively to the outer edges where gradient magnitudes are largest. More advanced approaches to identifying and plotting the isophotes are possible; however, this simple example highlights the additional meaning that can be extracted from the original data using a control grid constraint to overcome rather than eliminate noise.

A.   B.

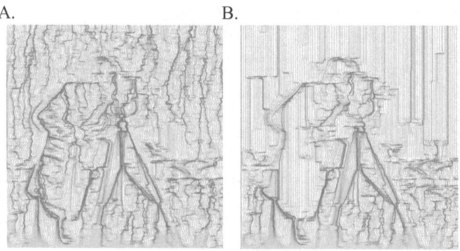

Figure 3.8: Both image panes show approximate isophote sketches of the cameraman image using displacements computed with the CGI approach. (A.) Results from using raw image derivatives. (B.) Results using image derivatives where derivatives with magnitudes less than 1% of the maximum image intensity have been set to zero and associated displacements have been voided.

Program 3.2: Compute displacement pairs for each pixel in an image along with a CGI estimate of the gradient angles

```
1   function [d1,d2,...%the computed displacements
2            phi]... % the gradient angle
3            = displacementsAndAngles(testImage,... %input image
4                          k,... %node spacing(s)
5                          clean,... %clip out small ...
                               derivatives
6                          visualize)   %show the images
7
8   %Computes the displacments for matching between rows and columns ...
9   %similar to linePair_timeWarp but for multiple lines of data and ...
        two sets
```

```
10  %of displacements.
11  %As with scharrDerivatives:
12  %optional clean arguement removes small derivatives,
13  %   if clean = 0 or empty then values are untouched, otherwise ...
       derivatives
14  %   with magnitudes less than clean*max(testImage) are set to zero.
15  %visualize generates a plot of the image, derivatives, and ...
       gradient angle
16
17  %uses the Scharr derivitive function and can generate a similar ...
       figure
18  if(nargin<4)
19      visualize = 0; %default is not to show figures
20      if(nargin<3)
21          clean = 0; %default is to not clean
22          if(nargin<2)
23              k=6; %default node spacing;
24              if(nargin<1)
25                  testImage = imread('cameraman.tif');%run as demo
26                  visualize = 1; %show figures for demo
27              end
28          end
29      end
30  end
31
32
33  %from the input define and adjust:
34  [M,N,colorChannels] = size(testImage);%line length
35  %for simplicity - run on grayscale only:
36  if(colorChannels>1)
37      testImage = rgb2gray(testImage);
38  end
39
40  %define the derivative images:
41  [Ix,Iy] = scharrDerivatives(testImage,clean,visualize);
42
43  %define the Interp matrix and nodes:
44  [interp,nodes] = defineInterpMatrix(N,k);
45  L = length(nodes);
46  %force the first and last displacements to be zero:
47  interp(:,1) = 0;
48  interp(:,end) = 0;
49
50  %step through column-by-column to find d1:
51  %parfor(n=1:N) %use the parallel toolbox
52  for(n=1:N) %otherwise
53      J = diag(Ix(:,n))*interp;
54      JtransJ = J'*J;
55      RHS = J'*(-1*Iy(:,n));
56
57      %solve for the displacments at the grid points:
58      %split the matrix into the diagonal components:
59      a = [0 JtransJ(2:L+1:end)];
60      b = JtransJ(1:L+1:end);
61      c = [JtransJ(L+1:L+1:end) 0];
62      %note that e = RHS;
```

```matlab
63      nodalDisps = TDMA(a,b,c,RHS);
64      d1(:,n) = interp*nodalDisps';
65  end
66
67  %define the Interp matrix and nodes:
68  [interp,nodes] = defineInterpMatrix(M,k);
69  L = length(nodes);
70  %force the first and last displacements to be zero:
71  interp(:,1) = 0;
72  interp(:,end) = 0;
73
74  %step through row-by-row to find d2:
75  %parfor(m=1:M) %use the parallel toolbox
76  for(m=1:M) %otherwise
77      J = diag(Iy(m,:))*interp;
78      JtransJ = J'*J;
79      RHS = J'*(-1*Ix(m,:)');
80
81      %solve for the displacments at the grid points:
82      %split the matrix into the diagonal components:
83      a = [0 JtransJ(2:L+1:end)];
84      b = JtransJ(1:L+1:end);
85      c = [JtransJ(L+1:L+1:end) 0];
86      %note that e = RHS;
87      nodalDisps = TDMA(a,b,c,RHS);
88      d2(m,:) = interp*nodalDisps';
89  end
90
91  %generate preliminary angle maps:
92  phiD1 = atand(d1);
93  phiD2 = atand(d2.^-1);
94  phi = 0.5*(phiD1+phiD2);
95  %note that when d1=d2=0 phi is 45 - for consistency, set to 0
96  phi = phi - phi.*(d1==0).*(d2==0);
97
98  if(visualize)
99   figure('Name','CGI Displacement Images','Color',[1 1 1]);
100   subplot('position',[0.03 0.5 0.4 0.4]);
101   imshow(repmat(double(testImage),[1,1,3])/double(max(testImage(:))),[])
102   title('Input Image','FontSize',14,'FontName','Times')
103   subplot('position',[0.5 0.5 0.4 0.4]);
104   imshow(phi,[-90,90])
105   colorbar('FontSize',14,'FontName','Times')
106   title('Approximated Gradient Angle (Degrees)',...
107       'FontSize',14,'FontName','Times')
108   subplot('position',[0.05 0.02 0.4 0.4]);
109   imshow(d2,quantile([d1(:);d2(:)],[.05 .95])) %exclude outliers
110   colorbar('FontSize',14,'FontName','Times')
111   title('Horizontal Displacements','FontSize',14,'FontName','Times')
112   subplot('position',[0.5 0.02 0.4 0.4]);
113   imshow(d1,quantile([d1(:);d2(:)],[.05 .95])) %exclude outliers
114   colorbar('FontSize',14,'FontName','Times')
115   title('Vertical Displacements','FontSize',14,'FontName','Times')
116   colormap('jet')
117  end
```

3.2   REGISTRATION OF TWO-DIMENSIONAL DATA: INTER-IMAGE REGISTRATION

In the previous section we looked at problems where a one-dimensional control grid formulation improved solution quality and made optimization more robust to noise. In this section we look to two-dimensional problems where the control grid framework serves as an essential constraint to make underdetermined optimization problems tractable. After briefly revisiting CGI for motion estimation, we will introduce two specific applications: mitigation of atmospheric turbulence distortion in video and medical image registration.

3.2.1   MOTION ESTIMATION

In Chapter 1 we covered many basics of motion estimation and in Chapter 2 we introduced the CGI approach to optical flow in the context of motion estimation. The fundamental equalities and error functions have been described, as well as a matrix formulation that allows for an analytical solution for the full system. Recall that the block-tridiagonal structure of the coefficient matrix involved in the optimization problem makes an analytical approach to the matrix inversion computationally feasible. In some cases, depending on image size and node spacing, the analytical approach can be faster than iterative methods [Frakes et al., 2013, Zwart et al., 2012]. Iterative approaches offer the opportunity to refine and adjust local solutions before incorporating them into the global framework [Frakes et al., 2008]. An adaptive approach (analytical or otherwise) allows for progressively localized refinement of solutions on a denser grid [Frakes et al., 2003]. Alternatively, the same node spacing can be utilized beginning with down-sampled or smoothed images before progressing to the original resolution data [Frakes et al., 2008]. In general, the control grid constraint can be incorporated into any optical flow solution framework. CGI can even be used *after* a dense motion field is calculated as a smoothing operation [Altunbasak and Tekalp, 1997].

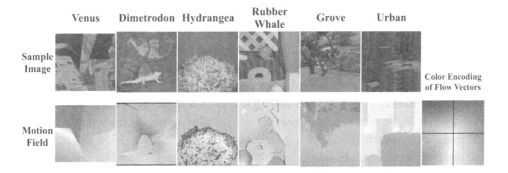

Figure 3.9:   Samples of images and motion fields taken from the Middlebury database.

In this section we compare the accuracy of motion field estimates relative to ground truth using a global/analytic CGI approach, a sliding-window/iterative CGI approach, block matching,

Table 3.1: Motion field accuracy and computation time are reported for several sample data sets from the Middlebury collection [Baker et al., 2011, Sun et al., 2010]. Lowest errors and times are indicated in bold.

		Horn-Schunck Optical Flow	Iterative CGI	Exhaustive Search Block Matching	Analytical CGI
	Duration	29.53	23.31	9.39	**2.73**
Venus	AAE	**5.49**	15.97	17.30	30.00
	Avg EPE	**0.34**	1.32	1.52	2.28
	Duration	43.40	27.35	12.29	**4.04**
Dimetrodon	AAE	**4.56**	5.39	14.09	9.43
	Avg EPE	**0.22**	0.30	0.87	0.48
	Duration	44.62	27.84	12.44	**4.02**
Hydrangea	AAE	**2.21**	5.16	6.72	16.13
	Avg EPE	**0.19**	0.50	0.54	1.51
	Duration	43.82	27.05	12.29	**4.05**
Rubber Whale	AAE	**3.80**	9.03	11.65	9.20
	Avg EPE	**0.12**	0.29	0.47	0.30
	Duration	66.14	40.50	17.05	**5.91**
Grove	AAE	**2.85**	6.56	11.40	32.50
	Avg EPE	**0.20**	0.48	0.90	1.58
	Duration	60.26	40.54	16.79	**6.06**
Urban	AAE	**4.08**	12.28	35.07	38.44
	Avg EPE	**0.46**	1.98	6.56	7.24

and optical flow to define motion field. The image pairs and motion fields (shown in Figure 3.9) along with the optical flow code are described in Baker et al. [2011] and available online from Sun et al. [2010]. Program 1.2 is used for block matching. The sliding-window adaptive CGI approach is described by Frakes et al. [2008]. The analytical approach is the direct implementation described in this book; that is, the matrix formulation for the globally connected control grid problem is solved via direct matrix inversion and no adjustments are made to discard low quality matches, prefer small displacements, or break up the control grid to better accommodate separate structures. For the iterative CGI approach, the image resolution is progressively increased from one-eighth to the original with a block side length equal to eight for all resolutions. For the analytical approach, the block size is progressively halved from $k = 64$ to $k = 8$. Table 3.1 details the computation times and reports the quantitative accuracy of the motion fields produced by each method in terms of the average angular error (AAE) and the average end point error (EEP).

3.2.2 MITIGATION OF ATMOSPHERIC TURBULENCE DISTORTION

To this point we have described two-dimensional CGI approaches to the brightness constraint (optical flow) primarily for motion detection. Atmospheric turbulence refers to a highly complex distortion process involving multiple atmospheric parameters including changes in refractive index, optical turbulence, and aerosol effects [Zwart et al., 2012]. The visual result of these distortions is perhaps most recognizable looking out over hot tarmac on an airplane runway. Mathematically, the distortion can be thought of as blurring, artificial displacement, and noise [Frakes et al., 2001]. The artificial displacement gives the illusion that things are moving in a quasi-periodic way, oscillating in and out of true position while noise and blurring detracts from image quality and complicates registration. The three part, image-based model of atmospheric distortion can be expressed in simplified form as:

$$\mathbf{I}(x, y, t) = a\{\mathbf{F}(x, y, t) * \mathbf{H}(x, y, t)\} + \boldsymbol{\eta}(x, y, t) \tag{3.16}$$

where $\mathbf{I}(x, y, t)$ is a measured intensity value representing the true brightness $\mathbf{F}(x, y, t)$. The scene is captured at time t, $\mathbf{H}(x, y, t)$ represents the local blurring function, a represents an artificial displacement operator, and $\boldsymbol{\eta}(x, y, t)$ represents additive noise. As in previous sections, we will migrate from x, y, t to m, n, p as we shift to discrete discussions. Reference to this model as image-based distinguishes it from physical models or adaptive optics compensation and places it in the same general category as blind deconvolution and lucky-region fusion approaches.

The model described in Equation (3.16) has been in use for over a decade to describe the atmospheric distortion problem. Generally speaking, work that builds upon this framework can be categorized as placing emphasis on deblurring (as in Li et al. [2005] and Mao and Gilles [2012]) or displacement compensation (e.g., Frakes et al. [2001]). Many of the displacement-focused works have used a CGI approach to optical flow to define the artificial displacements. In general, motion detection in the context of atmospheric turbulence correction is confronted with two main challenges: 1) to estimate the spatially local displacements efficiently (in real time) and accurately (with high sub-pixel resolution); and 2) to reconstruct distortion-compensated image frames with pristine quality while preserving legitimate motions. While these challenges may be less of a concern in more scientific applications (e.g., stellar recordings) where processing times are less restricted and legitimate motions are slow, applications to surveillance and defense (e.g., reading the numbers off of a moving vehicle in a hot climate) require that the challenges be overcome. While any of the previously described motion detection methods could be used, algorithms using a CGI framework are appealing given the time constraints imposed by this type of application.

Before showing results for atmospheric turbulence correction using both an analytical and an iterative CGI implementation, we briefly describe the other image processing steps and introduce the concept of registering to a reference frame. While extending previous discussions of registering a source and target frame to registering a reference and target frame are straightforward, considerations regarding the generation of the reference frame and direction of the registration are noteworthy.

Prior to motion estimation, the data are deblurred and the reference frame is generated. Simple unsharp masking and temporal averaging can be employed to accomplish these steps. Preliminary, rigid-body motion correction may be necessary to align frames when camera motion is present (e.g., in unmanned aerial vehicle surveillance). For the method described here, the output of these preliminary processing steps is considered the reference frame and is used in the motion correction process to further correct for artificial displacements. Figure 3.10 shows the basic image processing work flow associated with correcting for the atmospheric distortions described in 3.16.

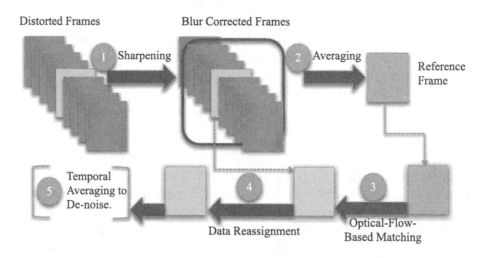

Figure 3.10: The image processing pipeline applied to correct for atmospheric distortion. (1) Blurring is addressed by a deblurring or sharpening operation. (2) The reference frame is generated using a temporal average of the deblurred frames. (3) Motion estimation is used to register the reference and current, deblurred frame. (4) The registration transform is used to place artificially displaced pixels in their proper locations. (5) Noise is mitigated using a temporal average of the motion corrected frames.

The advantage of the displacement-corrected frame over the reference frame is two-fold. Direct averaging of multiple, artificially displaced frames will introduce blur; this is an additional, post-processing-generated blurring beyond what is described in Equation 3.16 and is created rather than corrected for by a direct averaging step. While the artificial displacements can be considered periodic (meaning the average displacement over time is near zero), the average intensity at any given pixel location has no guaranteed relationship with the true value. As other objects displace in and out of a given pixel, they introduce blurring into the temporal average image. Furthermore, time averaging reduces temporal resolution and distorts moving objects causing ghosting or streaking artifacts. Using a registration transform to reposition the data in the current frame restricts the available intensities to those found in the current frame. For example, if the frame includes a moving vehicle, only the current 'copy' of the vehicle is available in generating the output image. A time average will contain multiple copies of the vehicle across a range of positions. The procedure for defining the registration transform is described next.

For a given pixel (m, n) in the reference frame, we define the 'true' data values from a deblurred (sharpened) and denoised (median filtered) frame, $\mathbf{I}$. We use the parameters d_1 and d_2 to describe the vertical and horizontal offsets to the 'true' data and apply the brightness constraint to assert that the intensity of the data is preserved as the location is artificially displaced:

$$\mathbf{R}(m, n, p) = \mathbf{I}(m + d_1, n + d_2, p). \tag{3.17}$$

As before, the Taylor series expansion can be employed to separate the offset and intensity variables:

$$\mathbf{R}(m, n, p) = \mathbf{I}(m, n, p) + \mathbf{D_1}(m, n, p)\frac{\partial \mathbf{I}(m, n, p)}{\partial x} + \mathbf{D_2}(m, n, p)\frac{\partial \mathbf{I}(m, n, p)}{\partial y}. \tag{3.18}$$

If we define $\bar{\mathbf{I}}(p)$ as the column vector of all pixel intensities from frame p (taken column-wise) and $\bar{\mathbf{d}}_1(p)$ and $\bar{\mathbf{d}}_1(p)$ as the vectors of displacements, then, abbreviating the vectors of the partial derivatives as $\bar{\mathbf{I}}_{\mathbf{x}}(p)$ and $\bar{\mathbf{I}}_{\mathbf{y}}(p)$, we can express Equation (3.18) for a full image and reference frame pair as:

$$\bar{R}(p) = \bar{\mathbf{I}}(p) + \text{diag}(\bar{\mathbf{I}}_{\mathbf{x}}(p))\bar{\mathbf{d}}_2(p) + \text{diag}(\bar{\mathbf{I}}_{\mathbf{y}}(p))\bar{\mathbf{d}}_2(p). \tag{3.19}$$

The two-dimensional control grid approach and the associated interpolating matrix Θ allow the full offset vectors to be defined from the control points or nodes, as:

$$\bar{\mathbf{d}}_1(m) = \Theta\bar{\mathbf{d}}_{1\mathbf{L}}(z), \tag{3.20}$$

and

$$\bar{\mathbf{d}}_1(m) = \Theta\bar{\mathbf{d}}_{1\mathbf{L}}(z). \tag{3.21}$$

Rewriting Equation (3.19) using the control points and introducing matrices $\mathbf{J_x}$ and $\mathbf{J_y}$ yields

$$\bar{R}(p) = \bar{\mathbf{I}}(p) + \bar{\mathbf{J}}_{\mathbf{x}}(p)\bar{\mathbf{d}}_{1\mathbf{L}}(p) + \bar{\mathbf{J}}_{\mathbf{y}}(p)\bar{\mathbf{d}}_{2\mathbf{L}}(p), \tag{3.22}$$

where

$$\mathbf{J_x}(p) = \text{diag}(\bar{\mathbf{I}}_{\mathbf{x}}(p))\Theta, \tag{3.23}$$

and

$$\mathbf{J_y}(p) = \text{diag}(\bar{\mathbf{I}}_{\mathbf{y}}(p))\Theta. \tag{3.24}$$

The problem of defining the offset variables describing the movement occurring between original frame $\mathbf{I}$ and reference frame $\mathbf{R}$ can then be addressed using a least squares approach; solving for the $\bar{\mathbf{d}}_{1\mathbf{L}}$ and $\bar{\mathbf{d}}_{2\mathbf{L}}$ vectors that minimize:

$$E(\bar{\mathbf{d}}_{1\mathbf{L}}, \bar{\mathbf{d}}_{2\mathbf{L}}) = [\bar{\mathbf{I}} - \bar{\mathbf{R}} + \bar{\mathbf{J}}_{\mathbf{x}}\bar{\mathbf{d}}_{1\mathbf{L}} + \bar{\mathbf{J}}_{\mathbf{y}}\bar{\mathbf{d}}_{2\mathbf{L}}]^2. \tag{3.25}$$

The indexing variable has been dropped from the notation for brevity; all values are for the same time frame. As detailed in Chapter 2, we can consolidate Equation 3.25 into a compact form with a block-tridiagonal coefficient matrix:

$$\mathbf{J}^T\bar{\Delta} = \mathbf{J}^T\mathbf{J}\bar{\mathbf{d}}_{\mathbf{L}}, \tag{3.26}$$

where

$$\bar{\Delta} = \bar{R} - \bar{I}, \tag{3.27}$$

and J and $\bar{d}_L$ are concatenated versions of J_x and J_y and $\bar{d}_{1L}\ \bar{d}_{2L}$, respectively, as detailed in Equation 2.22 and the associated material. Again, any one of a number of methods can be applied to solve for the displacements in Equation 3.26. Given the time restrictions native to surveillance applications (and the scope of this book) we will demonstrate results using both analytical and iterative CGI-based approaches.

Once the displacement vectors have been determined, constructing the corrected output frame involves reassigning the data from the corrupted time-frame to the corrected locations. For $H(m, n, p)$ an intensity in the corrected output image associated with input image $I(p)$, the originally collected data is resampled such that:

$$H(m, n, p) = I(m + D_1(m, n, p), n + D_2(m, n, p), p). \tag{3.28}$$

Determining the appropriate intensity will generally involve interpolation in the source frame, but this approach avoids costly regridding procedures necessitated by performing the registration in the opposite direction. Castleman [1979] refers to the approach as pixel filling as opposed to pixel carryover (p. 112).

A sample corrected video frame (along with the unprocessed source data) is shown in Figure 3.11 along with an enlargement of a stationary portion of the scene. Figure 3.12 highlights the quantitative performance of CGI-based displacement correction approaches for mitigating atmospheric turbulence artifacts, based on real-world, turbulence-distorted images of the static portion. In the left section, (artificial) changes in the intensity of a stationary portion of the video sequence are quantified for various correction approaches. For all image columns, the top-most row refers to unfiltered data from the original video sequence. The bottom rows all present results that incorporate simple sharpening and median filters as well as time averaging. In addition to the linear filters used for the second row of images, the bottom three rows highlight results obtained using a displacement correcting step that is either iterative CGI-based (third row) or based on an analytical approach (bottom two rows). In both the third and fourth rows a multiresolution approach is taken. Input images are down-sampled and the displacements are estimated in multiple passes as their resolution is returned to that of the original frame. The bottom row shows the results for directly computing the displacement field using the full resolution data. For all rows shown, the left-most column shows a subregion of the output, the middle column shows the total absolute difference in each pixel's intensity accumulated over a one hundred frame interval (relative to the first frame in the interval), and the right-most column shows the range of pixel intensities observed over the interval. The bottom right section plots the absolute differences in pixel intensity for the entire subregion over time (i.e., the images at right are summed across time and the plot at left collapses the data in space and displays them across time). Results for the iterative and analytical approaches are similar in quality and most notably differentiated by the computation times required. The multiresolution analytical compensation approach takes on average 50% the time required for the iterative version (the direct

Original   Corrected

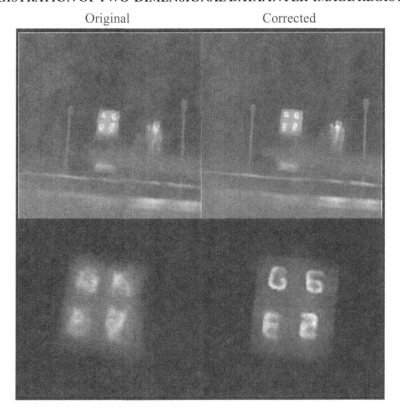

Figure 3.11: Qualitative examples of uncorrected and corrected video frames along with an enlargement of a small, stationary section of the scene.

implementation requires just 25%) when the final control grid resolution is set to 8×8 blocks. Given the volume of data involved with video processing and the time demands associated with many applications (especially surveillance), this faster approach to artificial displacement correction makes it more practical to address the apparent motion introduced by atmospheric turbulence.

3.2.3   MEDICAL IMAGE REGISTRATION

In this final subsection, we will introduce some of the special considerations relevant to the registration of medical images. Specifically, we explore registering two images of the same (person's) anatomy imaged with different modalities or contrasts. We focus on this application because it directly violates the brightness constraint we've built all previous approaches on.

Registration in medical imaging plays a critical role in a wide range of applications [Fischer and Modersitzki, 2008, Klein et al., 2009]. Images are frequently registered to a

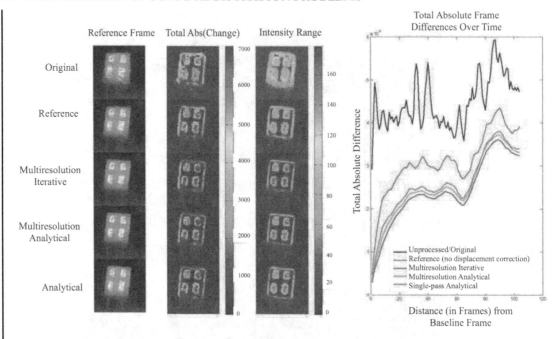

Figure 3.12: Examples of results from various correction procedures applied to a small static region from a sample data set affected by atmospheric turbulence. The input and output frames are shown along with quantitative plots showing the intensity changes in what should be a static region of the image.

generic template for comparison to 'normal' anatomy, atlas-based segmentation, or use in research studies. Multiple images of the same anatomy may be collected at different times, before and after a medical treatment, with and without a contrast agent, using magnetic resonance image (MRI) with different contrasts, or across different imaging modalities including computed axial tomography (CAT) X-ray scans. Many approaches for rigid and non-rigid registration have been developed to accommodate some or all of these registration scenarios. Our focus is on extending the previously described capabilities of the CGI implementation of optical flow to accommodate situations where the general content (shapes) is constant, but orientations and relative intensities are subject to change.

As an example, Figure 3.13 shows idealized examples of two MRI images collected with different contrast parameters. The left most image shows what is referred to as T1 contrast and the middle image shows T2 contrast. The images are synthetic and perfectly aligned (described in Collins et al. [1998] and available from http://brainweb.bic.mni.mcgill.ca/brainweb/). Also shown (far right) is the contrast ratio between the two images. Unlike global brightness changes resulting from, for example, adjustments to lighting conditions, changes to contrast parameters in medical imaging impact the relative contrast ratio. Some tissues may shift from bright to dark while others shift from dark to light, and some regions make experience very little change. As such, even for perfectly aligned images like those shown in Figure 3.13, the T1 to T2 intensity ratio is spatially varying.

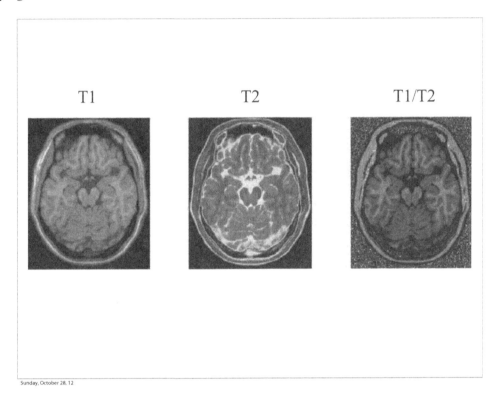

Figure 3.13: Synthetic brain MRI data representing T1 and T2 contrast along with the contrast ratio map.

To explore how the CGI approach to optical flow can be applied in the context of an image pair with a spatially varying contrast ratio, we formulate the intensity relationship between two aligned images $\hat{A}$ and $\hat{B}$ using the contrast ratio map $\mathbf{C}$ such that:

$$\hat{\mathbf{A}}(x, y) = \mathbf{C}(x, y) \cdot \hat{\mathbf{B}}(x, y). \tag{3.29}$$

In this framework, the original images $\mathbf{A}$ and $\mathbf{B}$ are registered to a mutual coordinate system where $\mathbf{C}$ is defined. Referring to the symmetric framework described in Figure 2.4, the coordinate system of $\mathbf{C}$ defines the middle position and we assert that the coordinates in $\mathbf{A}$ and $\mathbf{B}$ are symmetrically displaced such that:

$$\hat{\mathbf{A}}(m, n) = \mathbf{A}(\hat{m}, \hat{n}) = \mathbf{A}(m - \mathbf{D_1}(m, n), n - \mathbf{D_2}(m, n)), \tag{3.30}$$

and
$$\hat{B}(x, y) = B(\hat{m}, \hat{n}) = B(m + D_1(m, n), n + D_2(m, n)). \tag{3.31}$$

In contrast to the original optical flow framework, the displacements as well as the matrix C are unknown. Despite this significant change, we can still construct an optimization problem for which CGI is well-suited and that follows the general structure associated with optical flow and motion estimation.

Equation 3.29 written in terms of the original (unregistered) data is:

$$A(m - D_1(m, n), n - D_2(m, n)) = C(m, n) \cdot B(m + D_1(m, n), n + D_2(m, n)). \tag{3.32}$$

Using the Taylor series expansion to separate the displacement and coordinate terms yields:

$$A - A_x \circ D_1 - A_y \circ D_2 = C \circ B + C \circ B_x \circ D_1 + C \circ B_y \circ D_2, \tag{3.33}$$

where $C \circ B$ indicates the Hadamard product or element-by-element multiplication and the equality is extended for the full matrices. Rearranging yields:

$$A - C \circ B = D_1 \circ [A_x + C \circ B_x] + D_2 \circ [A_y + C \circ B_y], \tag{3.34}$$

which is very similar to the equation generated for the standard two-dimensional registration; however, C remains unknown.

Using an iterative framework to estimate C where $C^{i=0}(m, n) = A(m, n)/B(m, n)$ and $C^{i+1}(m, n) = \hat{A}^i(m, n)/\hat{B}^i(m, n)$ allows us to derive the error function for the current displacement estimates as:

$$E(D_1^i, D_2^i) = \left(D_1^i \circ [\hat{A}_x^i + C^{i-1} \circ \hat{B}_x^i] + D_2^i \circ [\hat{A}_y^i + C^{i-1} \circ \hat{B}_y^i] \right)^2. \tag{3.35}$$

The estimates for the registered images are constructed from the original data using the sum of displacements from all iterations such that:

$$\hat{A}^n(x, y) = A(x - \sum_{i=1}^{n} D_1^i(x, y), y - \sum_{i=1}^{n} D_1^i(x, y)), \tag{3.36}$$

and

$$\hat{B}^n(x, y) = B(x + \sum_{i=1}^{n} D_1^i(x, y), y + \sum_{i=1}^{n} D_1^i(x, y)). \tag{3.37}$$

All derivatives are computed using the current, registered input images. When using a control grid framework to optimize the displacements, the grid resolution can be refined in tandem with the estimates for the contrast ratio image, C. In this way the initial estimates for C impact a broader region with the expectation that the preliminary estimates are on average accurate. As the image registration improves, the ratios are pooled over a smaller region and subject to lesser averaging effects.

3.3   SUMMARY

In this chapter, we extended our coverage of CGI optimization beyond traditional optical flow and motion estimation. We used the framework for CGI optimization of the optical flow brightness constraint to register both one- and two-dimensional signals separated in time based on matches in intensity. We also considered applications involving signals separated in space where the brightness constraint amounts to isophote identification. In the final application, we addressed registration problems that cannot be handled directly using assumptions about constant intensity, by incorporating a function describing the intensity relationship between images in a pair. In all of the applications, the registration, matching, or alignment of signals was the primary purpose of the CGI optimization. Using CGI to determine registration transforms and in turn to define new, uncollected data is the focus of the next chapter.

CHAPTER 4

Application of CGI to Interpolation Problems

In the previous chapters, we established that the foundation of optical flow and any application of the brightness constraint is identifying a one-to-one match between two datasets based on intensity. This intensity-based registration leaves us with a motion or displacement field describing the vectors connecting the matches in the two datasets. In this chapter we explore interpolation methods that comprise inserting new, estimated data along those vectors.

4.1 INTERPOLATION OF 1D DATA: INTER-VECTOR INTERPOLATION

In Section 3.1.2 we introduced CGI implementations of the brightness constraint as a way to identify isophotes within an image. Using the local orientation of the isophotes, we can define interpolation kernels that better accommodate the local image structure than traditional, square, or radial kernel shapes. Fundamentally, interpolation involves identifying a neighborhood of related pixels and estimating the value of additional points within the neighborhood based on known values. While traditional, fixed-shape interpolation kernels incorporate a local neighborhood of pixels based strictly on spatial distance, adaptive kernel approaches use additional image information to modify the neighborhood and adjust it based on image structure. In this section we use local approximations of the image isophotes defined using one-dimensional CGI to shape the interpolation neighborhood. We apply this interpolation technique first to generic image enlargement (single-image super-resolution) and then to the specific, related problem of intra-frame video deinterlacing.

4.1.1 SINGLE-IMAGE SUPER-RESOLUTION

Single-image super-resolution (interpolation, zooming, upsampling, or resizing) artificially increases the pixel density for viewing or printing and has important applications in almost every area of digital imaging. As such, considerable research has been focused on developing interpolation algorithms for both general and specific purposes [Algazi et al., 1991, Allebach and Wong, 1996, Aly and Dubois, 2005, Asuni and Giachetti, 2008, Atkins et al., 2001, Celik and Tjahjadi, 2010, Cha and Kim, 2007, Giachetti and Asuni, 2008, Guo et al., 2012, Han et al., 2010, Jensen and Anastassiou, 1995, Lee et al., 2010, Lee and Yoon, 2010, Li and Orchard, 2001, Liu et al., 2011, Mallat and Yu, 2010, Manjón et al., 2010, Morse and Schwartzwald, 2001, Park and Jeong, 2010, Ramani et al., 2010,

Temizel, 2007, Wang and Ward, 2007, Zhang and Wu, 2008, Zwart and Frakes, 2011]. Despite the superior quality (quantitative and qualitative) of the results achievable with more advanced methods, interpolators based on approximations of the ideal sinc kernel (pixel replication, bilinear, bicubic, and higher-order splines) are still prevalent in many consumer applications because of their flexibility and speed. Many applications where the computational burden and restrictions (e.g., expansion by factors of two only) of adaptive methods are unacceptable suffer from the blurring, ringing, jagged edges, and unnatural isophotes common to fixed-kernel interpolation. Approaches that provide an acceptable trade-off are an area of active research [Mallat and Yu, 2010, Zhang and Wu, 2008, Zwart and Frakes, 2011]. Using the same fast, analytical approach to isophote approximation described in the previous chapter, we have developed an efficient approach to interpolation that produces results with visual quality and quantitative accuracy on par with state-of-the-art adaptive kernel approaches, and with greater flexibility and speed. Our approach, based on a control grid implementation of the one-dimensional brightness constraint, has been termed segment adaptive gradient angle (SAGA) interpolation [Zwart and Frakes, 2012]. The approach is similar to the parallelogram shaped kernel approach detailed by Wang and Ward [2007]; however, in place of special cases, SAGA uses control nodes to define a regularized or segmented framework. By interpolating along isophotes rather than along the image lattice, SAGA effectively adapts to image structure and aligns the axis of the interpolation kernel along the gradient angle.

In describing the CGI approach to isophote detection, we have already established many of the mathematics involved in SAGA interpolation. We again approximate the isophote locally with a line of constant intensity such that:

$$I(m, n) = I(m + d_1, n + 1), \tag{4.1}$$

and

$$I(m, n) = I(m + 1, n + d_2). \tag{4.2}$$

The SAGA method comprises using a CGI framework for determining the isophote approximating vectors $[\pm d_1, \pm 1]$ and $[\pm 1, \pm d_2]$ and interpolating at intermediate locations along these vectors. Figure 4.1 extends the one-dimensional displacement framework placing an interpolated data point along the connecting vector. In this example case, matches are made from pixels in one row ($x = m$) to locations in the row below ($x = m + 1$) and intermediate locations are determined using the horizontal displacements:

$$I(m + \Delta_m, n + d_2\Delta_m) = (1 - \Delta_m)I(m, n) + (\Delta_m)I(m + 1, n + d_2), \tag{4.3}$$

with $0 \leq \Delta_m \leq 1$. This process implies a continuous function for defining intensities along the isophote approximating vector and provides the foundation for the SAGA algorithm. Constructing the higher resolution output image requires additional interpolation and convolution gridding steps using traditional, sinc-based kernels.

The complete set of displacements (one horizontal and one vertical for each pixel) can be used to define four 'matched' locations with respect to the original image. For the pixel at location (m, n)

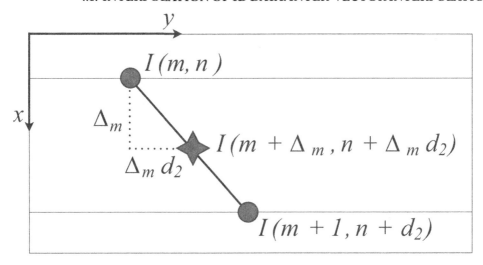

Figure 4.1: The displacement vector $[1, d_2]$ defines a connection between one node and a location in the next row. New data are calculated along the displacement vector by linear interpolation.

these locations are $(m \pm \mathbf{d}_1(m, n), n \pm 1)$ and $(m \pm 1, n \pm \mathbf{d}_2(m, n))$. Because the displacements are frequently non-integer, the destinations or end points are generally off-grid necessitating an intermediate interpolation step to define the intensity at the matched location. The interpolated intensity at the matched location is then used to define the new data along the displacement vectors using linear interpolation (as in Equation 4.3). Intensities in the high resolution image are defined based on original intensity data in the low resolution image and the interpolated intensities following convolution gridding to the high resolution lattice. Figure 4.2 shows a schematic of this process and contrasts it with bilinear interpolation. While Figure 4.2 shows expansion by a factor of two, the resolution of the output lattice is unrestricted and new data can be inserted at any prescribed density along the isophote approximations.

Importantly, each of the four sets of matching vectors results in a different output image that has been directionally interpolated along one axis. In the complete, uniformly enlarged image, columns consisting of entirely new data are directionally interpolated using the displacement vectors associated with d_1 (new rows of data are directionally interpolated using vectors associated with d_2). The dimension that is not directionally interpolated is expanded using a non-adaptive, one-dimensional interpolation or gridding kernel. Figure 4.3 depicts how the row and column approaches define data at different locations in the high-resolution, uniform grid. Each interpolated pixel in the high-resolution image is directionally determined in both of the row images and/or both of the column images. Details on how the four images are combined to form a single output image follow.

Thus far we have emphasized that in practice, four independently interpolated and distinct images are generated based on the four matching vectors identified for each pixel in the original

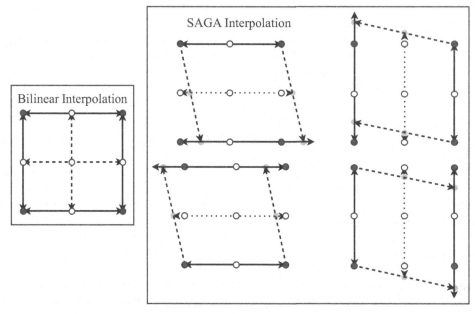

— Directly interpolated from existing data
- - - - Interpolated from interpolated (and existing) data
· · · · · · · Gridding of interpolated data
● Existing data
○ Interpolated, high resolution image data
● Interpolated, intermediate data (not included in final lattice)

Figure 4.2: The displacement vectors define connections between each pixel and a location in the next row. New data are calculated along the displacement vector by linear interpolation.

image. Ideally, all of the vectors approximating the isophote intersecting a given pixel are consistent such that the four estimates are equivalent; however, this is rarely the case and in combining the four interpolated images, a weighting system that emphasizes the best estimate is desirable. Based on the knowledge that well estimated isophotes will yield similar interpolant estimates, we propose estimate agreement as a measure of quality. Expressing the average estimate for the high resolution image (H) as:

$$\hat{\mathbf{H}} = \frac{1}{4} \sum_{i=1}^{4} \mathbf{H}_i, \tag{4.4}$$

the pixel-wise deviations of each high resolution estimate $(\mathbf{H}_i)$ from the average can be used to define its individual contribution to the final image. Specifically, using the squared difference as a

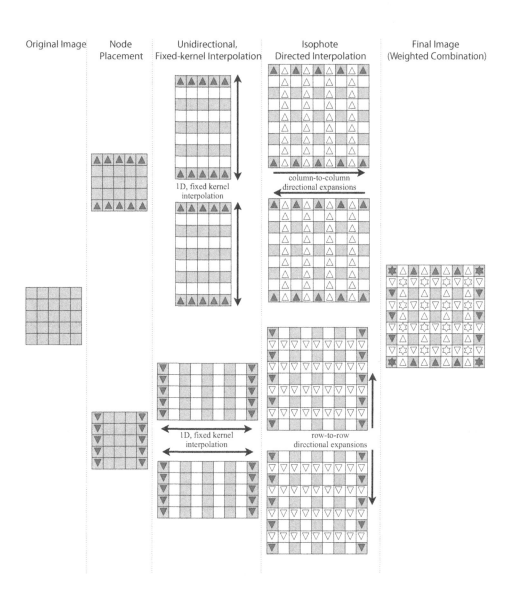

Figure 4.3: The final image is built from four independently interpolated images that are based on the four sets of matching vectors.

measure of deviation:

$$\mathbf{\Delta}_i(m, n) = |\hat{\mathbf{H}}(m, n) - \mathbf{H}_i(m, n)|^2, \tag{4.5}$$

the contribution of the individual estimates to the final image is adjusted according to:

$$\mathbf{H}(m, n) = \frac{\sum_{i=1}^{4} \mathbf{W}_i(m, n)\mathbf{H}_i(m, n)}{\sum_{i=1}^{4} \mathbf{W}_i(m, n)}, \tag{4.6}$$

where the weights $(\mathbf{W}_i(m, n), i \in 1, 2, 3, 4)$ at each pixel location are defined by some monotonic, decreasing function. For example:

$$\mathbf{W}_i(m, n) = \sum_{j=1}^{4} \mathbf{\Delta}_j(m, n) - \mathbf{\Delta}_i(m, n). \tag{4.7}$$

Alternatively, any of the high resolution images can be used as the final output with limited impact on visual quality. The main advantages of SAGA are its low complexity and flexibility making a single estimate approach an appealing way to deliver any even faster result.

Recalling that the least-squares optimization associated with the CGI approach to the one-dimensional brightness constraint involves tridiagonal coefficient matrices, the overall order of the SAGA approach in computing the displacement parameters is $O(MN)$ for an $M \times N$ image. Importantly, the complexity is related to the original image dimensions and not influenced by the enlargement factor. Furthermore, once the displacement parameters are computed they can be applied and reapplied for any enlargement factor. In addition to having low raw computational complexity, the SAGA algorithm is easily configured for parallelization as data can be independently processed one line at a time. Given the ever increasing range of image viewing devices and consumer expectations for interactivity, this feature is viewed as an important benefit of SAGA interpolation. Figure 4.4 provides an example of progressively enlarging a color image both with a SAGA approach and pixel replication.

4.1.2 VIDEO DEINTERLACING

A primary advantage of the SAGA interpolation approach is flexibility in accommodating arbitrary scaling factors. When such flexibility is not required, additional constraints can be imposed for efficiency gains. As a specific example, a symmetric approach to the one-dimensional brightness constraint applies well to repeatedly interpolating new rows of data between existing rows of interlaced video data. Interlaced video formats are still common despite the prevalence of digital display technologies. Interlacing capitalizes on the human visual system's relatively low temporal resolution to reduce transmission bandwidths. This is accomplished by transmitting or storing only half the data lines for each frame, and alternating the lines that are refreshed in each frame. Digital displays require data for every row of every frame, necessitating a deinterlacing step. Much like image resizing, a wide range of approaches to deinterlacing have been presented [De Haan and Bellers, 1998, Lee et al., 2000, Oh et al., 2000]; however, performance trade-offs and algorithm efficiency are perhaps even

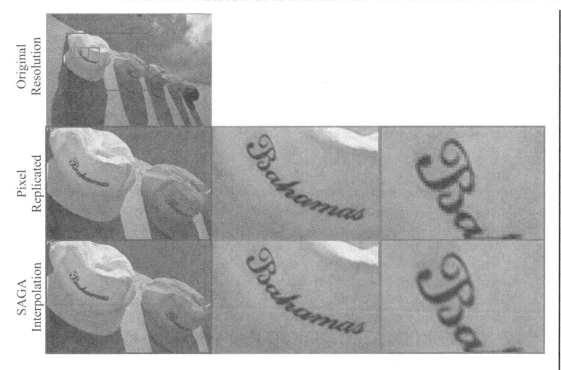

Figure 4.4: An adaptive SAGA approach to interpolation produces fewer jagged edges and artifacts in this example of progressive or dynamic zooming. Enlargement factors of 2, 6, and 12 are shown.

more relevant for deinterlacing applications. As a result, low complexity approaches such as line doubling and linear interpolation are still popular, and simple adaptive methods like edge-based line averaging (ELA) [Lee et al., 2000] are often preferred even in hardware implementations. Use of a CGI framework to define local isophote or edge orientation enables edge-based deinterlacing with higher angular resolution than ELA at a manageable computational cost.

In any deinterlacing approach, each newly generated row is bordered by two known rows in the same time frame and also two known rows in the neighboring time frames. Either pair can be used to directly estimate new data points using simple averaging or an edge-based approach. For the CGI-based approach, a row-to-row-next and/or row-to-row-previous framework can be used to link pixels in the known rows. Applying the symmetric brightness constraint to this framework yields:

$$I(m - 1, n - d, p) = I(m, n, p) = I(m + 1, n + d, p). \qquad (4.8)$$

The appropriate value for d at every pixel in the new row is determined as for the general case; however, the cost function is modified to form:

$$E(d) = \left[I(m + 1, n, p) - I(m - 1, n, p) + d \left(\frac{\partial I(m + 1, n, p)}{\partial y} + \frac{\partial I(m - 1, n, p)}{\partial y} \right) \right]^2.$$

$$(4.9)$$

The estimate for the new data point at horizontal position n in the new row positioned half-way between rows $m - 1$ and $m + 1$ is simply:

$$I(m, n, p) = 0.5(I(m - 1, n - d, p) + I(m + 1, n + d, p)).\qquad(4.10)$$

The appropriate values for $I(m - 1, n - d, p)$ and $I(m + 1, n + d, p)$ are interpolated from the existing lines of data using a one-dimensional, non-directional interpolator, which yields an intra-frame estimate of the full frame. Figure 4.5 highlights the image quality improvements relative to ELA. In contrast to the general interpolation approach, no new columns are interpolated (directionally or otherwise) and no gridding step is required for deinterlacing which results in a significantly reduced computational burden. Inter-frame data can also be incorporated as in the spatio-temporal variant of ELA (STELA) [Oh et al., 2000].

4.2   INTERPOLATION OF 2D DATA: INTER-IMAGE INTERPOLATION

Just as displacements approximating image isophotes can be used to direct and adapt interpolation kernels, displacement vector fields that describe motion between or register two images can be used to reshape the neighborhoods used in inter-image interpolation. Registration-based interpolation uses a registration transform, which maps one set of known data to a spatially or temporally related set, to generate super-resolution data along the mapping. The fundamental assumption of registration-based interpolation is that the data bounding interpolants have similar features that the algorithm can identify and link [Penney et al., 2004]. As a result, registration-based interpolation is often applied to interpolate between images in a volumetric data set or between frames of video [Frakes et al., 2008]. Between adjacent video frames, the connections represent motion vectors that link the time-varying locations of objects [Chen and Lorenz, 2012, Lim and Park, 2011]. We begin by exploring interpolation in the time dimension (inter-frame interpolation) and then extend our discussion to include interpolation of volumetric image data.

4.2.1   INTER-FRAME INTERPOLATION

While compression is a more ubiquitous application of motion estimation, many inter-frame interpolation strategies also rely on motion estimation. In a sense, compression leverages some additional information that allows a good approximation of the original data to be interpolated well based on a subset of those data. Motion-based compression defines 'missing' frames by using the motion field to relocate data in the reference frame and adding in a low bit-depth residual frame. Motion-based

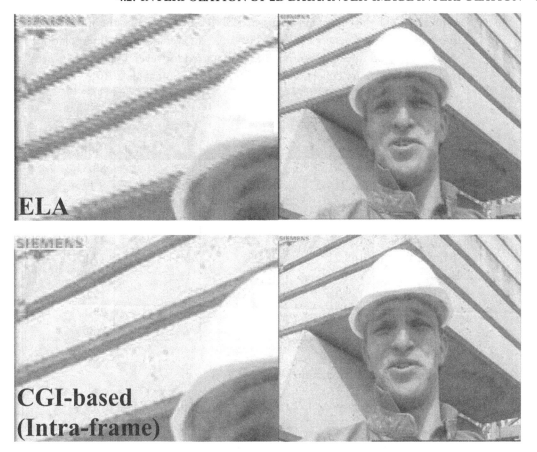

Figure 4.5: The edge angles on the left side of the frame are poorly addressed by the quantized edge angle approximations available using the ELA approach.

interpolation defines 'additional' frames by using a fractional version of the motion field and residual frame to define a new frame between the reference and target images. If the reference frame $\mathbf{I}(p)$ and the target frame $\mathbf{I}(p + 1)$ are linked by the motion field $[\mathbf{D_1}, \mathbf{D_2}]$ such that the residual frame, $\mathbf{R}(p + 1)$, is defined (pixel-wise) as:

$$\mathbf{R}(m + d_1, n + d_2, p + 1) = \mathbf{I}(m + d_1, n + d_2, p + 1) - \mathbf{I}(m, n, p), \text{ where}$$
$$d_1 = \mathbf{D_1}(m, n) \text{ and} \tag{4.11}$$
$$d_2 = \mathbf{D_2}(m, n),$$

Table 4.1: The accuracy (in terms of PSNR in decibels) of the inter-frame interpolation results generated using the frame pair average (linear interpolation) and motion field directed interpolation is reported for several sample data sets from the Middlebury collection [Baker et al., 2011, Sun et al., 2010].

	Linear	Analytical CGI	Iterative CGI	Horn-Schunck Optical Flow
Beanbags	107.41	88.87	88.75	**80.51**
Dimetrodon	29.21	10.1	**7.79**	8.19
Dog Dance	70.16	28.59	26.59	**21.78**
Grove	314.69	107.16	**52.12**	52.72
Hydrangea	103.5	33.7	**16.20**	17.60
Mini Cooper	168.21	98.05	21.40	**17.02**
Rubber Whale	7.19	5.59	11.57	**5.31**
Urban	139.41	105.0	19.59	**15.57**
Venus	200.36	89.50	35.38	**29.17**
Walking	18.05	9.7	9.45	**9.38**

then we can define an interpolated frame at time $p + \Delta_p$ as:

$$\mathbf{I}(m + \Delta_p d_1, n + \Delta_p d_2, p + \Delta_p) = \mathbf{I}(m, n, p) + \Delta_p \mathbf{R}(m + d_1, n + d_2, p + 1)$$
$$= \Delta_p \mathbf{I}(m + d_1, n + d_2, p + 1) + (1 - \Delta_p)\mathbf{I}(m, n, p)).$$

$$(4.12)$$

Effective compression is based on the assumption that the motion field can be defined such that:

$$|R(m, n, p + 1)| < |I(m, n, p + 1) - I(m, n, p)|. \qquad (4.13)$$

Similarly, the use of a motion field to direct inter-frame interpolation improves the estimate over (for example) a linear approach where:

$$\mathbf{I}(m, n, p + \Delta_p) = \Delta_p \mathbf{I}(m, n, p + 1) + (1 - \Delta_p)\mathbf{I}(m, n, p)). \qquad (4.14)$$

Table 4.1 illustrates this result for images from the Middlebury collection [Baker et al., 2011, Sun et al., 2010]. The accuracies of interpolation results obtained using iterative and analytical implementations of CGI-based optical flow are again compared to those achieved with an iterative approach based on the Horn-Schunck framework. As with motion fields themselves, the CGI framework generally offers a trade-off where accuracy is compromised for efficiency.

4.2.2 INTER-SLICE INTERPOLATION

The mathematics and algorithms described for inter-frame interpolation of times series data can be applied directly to inter-slice interpolation of volumetric data [Frakes et al., 2008]. Whereas the displacement vectors used for inter-frame interpolation are directly related to physical motion, the displacements determined using the brightness constraint in a volumetric context track isophotes and link isointense structures through multiple slices. Medical diagnostics is a common (volumetric imaging) scenario where the through-plane or inter-slice resolution is limited and interpolation may be desirable. Detailed in Frakes et al. [2008], inter-slice interpolation of medical (magnetic resonance) images is shown in Figure 4.6.

4.3 SUMMARY

In this chapter we have extended our discussion from the previous chapter to cover interpolation methods for both single images or frames and image stacks or volumes. In all cases we used the CGI approach to upholding the brightness constraint to define a pixel-by-pixel registration transform that links two locations within the collected data set. Having defined the intensities at matched locations, new data are placed along the registration vector that connects two locations, and the final image or volume lattice is defined by gridding the new data onto a regular pixel array. We also showed that for symmetric implementations, the gridding step can be avoided by defining the registration vectors to pass through the new high resolution locations.

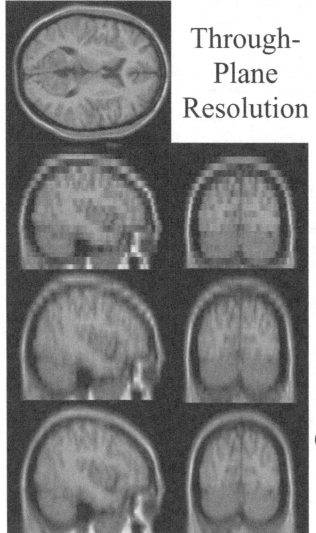

Through-Plane Resolution

Original Reformatted Images

Linear Interpolation

CGI, Brightness-Directed Interpolation

Figure 4.6: The in-plane resolution of the synthetic brain image is $1mm \times 1mm$ and the slice thickness is $9mm$. The alternate projections are shown along with images resampled to isotropic resolution using linear and the CGI-directed approach to inter-slice interpolation.

CHAPTER 5

Discussion and Conclusions

In this book we focused primarily on the use of control grids to minimize error functions related to the brightness constraint. We explored the registration process itself along with applications and extensions into interpolation. Prior to concluding we review some of the strengths and weaknesses of the control grid formulation along with potential avenues for future work.

5.1 STRENGTHS AND WEAKNESSES

The use of CGI to constrain an optimization problem has direct consequences that may be advantageous or problematic depending on the application. Furthermore, in implementing CGI, allowing modifications or exceptions to the underlying mathematics can provide a mechanism for overcoming some inherent limitations. In this section we will highlight some of the strengths and weaknesses of CGI-based methods before briefly discussing ways that certain weaknesses can be overcome.

In Chapter 1, we compared and contrasted two common motion estimation techniques: block matching and optical flow. The descriptions of motion defined using block matching algorithms are limited to uniform translations of entire blocks of pixels. This is a clear weakness in that the types of movement that can be characterized are severely restricted (e.g., rotation and zooming cannot be described), objects that span multiple blocks may be artificially split by the estimated motion, and objects smaller than the block size may be neglected. Block matching approaches always involve iterative minimization and require explicit evaluation of match quality. Exhaustive searches can become prohibitively expensive and many fast block matching approaches add additional assumptions (e.g., locally planar intensities) to decrease the number of candidate motion vectors.

Optical flow generates a dense motion field wherein the motion of each pixel is represented as a translation. Depending on the constraints placed on the estimated displacement vectors, a wide range of motions can be described including rotations and non-rigid-body deformations. Constraints on the motion field (particularly those related to its smoothness) can be imposed to promote solutions such that nearby pixels move in similar ways; however, objects are not explicitly segmented or kept together. The optical flow approaches we have described are based on the first order Taylor series expansion of the brightness constraint and make the assumption that intensities are locally planar.

In Chapter 2, we explored the use of a control grid to constrain the solutions to optical flow. CGI enforces a particular type of smoothness constraint on the estimated motion field and modifies the approach to resemble a hybrid of block matching and optical flow. One major distinction between a CGI approach and either optical flow or block matching is the global connectivity of the model. While other approaches promote global smoothness as part of the optimization, the CGI framework formally defines the local smoothness constraint. The connected framework and predetermined solution structure enable analytical approaches that provide a globally optimal solution. Real data, however, may not conform to the global connectivity enforced by the CGI framework.

In Chapter 3, we looked at approaches to signal registration using CGI. For cases such as dynamic time warping, a deformable model with global connectivity is a logical selection. Chapter 3 also showed that tracking a moving object in a static scene using a pure CGI framework can result in significant errors in the estimated motion field. A primary reason for such errors is that inaccurate motion estimates do not necessarily have consequence in the context of matching errors. Consider a plane moving across a clear sky. There is no penalty (in terms of the brightness constraint) associated with a motion field that displaces the surrounding sky along with the plane. Iterative approaches that start with no displacement as the initial guess are more likely to leave the background in its appropriate location. Refinements of the motion field that reset (to zero) displacements that do not improve the local matching error can also be used to improve on misestimations. Externally imposed constraints on maximum displacement can also improve the fidelity of motion estimates. An iterative CGI framework that implements such adjustments and is allowed to deviate from the globally optimized solution can yield results that are much closer in accuracy to pure optical flow, while retaining the improvements in computation time characteristic of CGI.

In Chapter 4, we used CGI-based registration to look at directional interpolation methods. Interpolation can be thought of as weighted averaging within a neighborhood to define new data points inside that neighborhood. When neighborhoods are defined based strictly on spatial relationships, the neighborhood average may have little in common with a new point's true value. If a neighborhood is specifically defined to contain points with similar values, estimating new points in that neighborhood based on the average will be more effective. Approaches to upholding the brightness constraint based on CGI provide a highly efficient means of establishing similarly intense neighborhoods, and form an effective foundation for interpolation in multiple dimensions. The counterpoint to those strengths is that the relatively large regions that nodal displacement parameters influence can disregard and in turn distort fine textural information.

Overall the strengths and weaknesses of a CGI approach to optical-flow-type optimization problems are generally related to accuracy and efficiency trade-offs. CGI problems can be directly solved quickly producing a globally connected solution. At the expense of more complicated implementations and greater computational costs, the control grid structure can be dynamically adjusted and the grid selectively broken to accommodate data features that do not adhere to the connectivity framework; however, fine textures and features with very limited spatial extent remain difficult to handle well with CGI.

5.2   APPLICATION TO HIGHER-DIMENSIONS AND MULTIVARIATE OPTIMIZATION

In Chapter 3, we used a CGI framework to register two medical image slices despite differences in contrast. Incorporating the contrast ratio as a third degree of freedom was addressed using an iterative refinement procedure. CGI is a general approach to constraining optimization and can be applied in contexts not addressed in this work (with higher degrees of freedom) or for data of higher dimensions. For practical applications related to medical image registration, it would be desirable to register volumetric data sets collected with different contrast characteristics. This type of application has been addressed using optical flow and could be constrained with a higher-dimension control grid structure (i.e., prisms in place of polygons). In addition to higher spatial dimensions, optical flow techniques have been applied to vector fields considering color and flow data as true vectors (as opposed to multiple, separate scalars). Researchers have built on the brightness constraint and incorporated mass and divergence constraints into similar frameworks. These applications represent opportunities for further development and deployment of control grid formulations.

5.3   FINAL THOUGHTS AND CONCLUSIONS

In this book we have provided a broad overview of motion estimation and other registration problems that can be addressed using an intensity preserving brightness constraint. We have specifically focused on the use of control grid interpolation to structure the solutions to these problems and in turn facilitate the optimization process. In the context of one-dimensional signals and uniquely determined systems, we explored CGI as a regularizer employed to overcome noise and spurious solution behavior resulting from outliers. For under-determined, two-dimensional scenarios, we used the CGI framework to constrain the optimization problem and make it tractable. Throughout this book we have focused on analytical approaches to CGI problems. We have also demonstrated that the optimal solution to the explicit problem does not always yield the best result for a given application. Future developments will focus on refinement procedures and on extensions to higher dimensions.

Bibliography

Algazi, V., G. Ford, and R. Potharlanka (1991, April). Directional interpolation of images based on visual properties and rank order filtering. In *Proc IEEE International Conference on Acoustics, Speech, and Signal Processing, (ICASSP)*, Volume 4, pp. 3005–3008. DOI: 10.1109/ICASSP.1991.151035 55

Allebach, J. and P. W. Wong (1996). Edge-directed interpolation. In *Proc. Int. Conf. Image Processing*, Volume 3, pp. 707–710. DOI: 10.1109/ICIP.1996.560768 55

Altunbasak, Y. and A. Tekalp (1997). Closed-form connectivity-preserving solutions for motion compensation using 2-d meshes. *Image Processing, IEEE Transactions on 6*(9), 1255–1269. DOI: 10.1109/83.623189 10, 43

Aly, H. A. and E. Dubois (2005, October). Image up-sampling using total-variation regularization with a new observation model. *IEEE Transactions on Image Processing 14*(10), 1647–1659. DOI: 10.1109/TIP.2005.851684 55

Amiaz, T., E. Lubetzky, and N. Kiryati (2007, September). Coarse to over-fine optical flow estimation. *Pattern Recognition 40*(9), 2496–2503. DOI: 10.1016/j.patcog.2006.09.011 2

Arkin, R. C. (2012, August). The role of mental rotations in primate-inspired robot navigation. *Cognitive processing 13*(Suppl 1), S83–7. DOI: 10.1007/s10339-012-0467-7 2

Asuni, N. and A. Giachetti (2008). Accuracy improvements and artifacts removal in edge-based image interpolation. In *Proc. Int. Conf. Computer Vision Theory and Applications.* 55

Atkins, C. B., C. A. Bouman, and J. P. Allebach (2001). Optimal image scaling using pixel classification. In *Proc. IEEE Int. Conf. Image Processing (ICIP)*, Volume 3, pp. 864–867. DOI: 10.1109/ICIP.2001.958257 55

Baker, S., D. Scharstein, J. Lewis, S. Roth, M. Black, and R. Szeliski (2011). A database and evaluation methodology for optical flow. *International Journal of Computer Vision 92*, 1–31. 10.1007/s11263-010-0390-2. DOI: 10.1109/ICCV.2007.4408903 44, 64

Bradski, G. and A. Kaehler (2008). *Learning OpenCV: Computer Vision with the OpenCV Library.* Sebastopol, CA: O'Reilly. 33

Brox, T., B. Rosenhahn, J. Gall, and D. Cremers (2010, March). Combined region and motion-based 3D tracking of rigid and articulated objects. *IEEE transactions on pattern analysis and machine intelligence 32*(3), 402–15. DOI: 10.1109/TPAMI.2009.32 2

Bruhn, A., J. Weickert, and C. Schnörr (2005). Lucas/Kanade meets Horn/Schunck: Combining local and global optic flow methods. *International Journal of Computer Vision 61*, 211–231. 10.1023/B:VISI.0000045324.43199.43. 7

Castleman, K. R. (1979). *Digital image processing*. Englewood Cliffs, N.J.: Prentice-Hall. Kenneth R. Castleman.; :ill. ;25 cm; Includes bibliographical references and index. 9, 48

Celik, T. and T. Tjahjadi (2010, July). Image Resolution Enhancement Using Dual-Tree Complex Wavelet Transform. *IEEE Geoscience and Remote Sensing Letters 7*(3), 554–557. DOI: 10.1109/LGRS.2010.2041324 55

Cha, Y. and S. Kim (2007, June). The error-amended sharp edge (EASE) scheme for image zooming. *IEEE Transactions on Image Processing 16*(6), 1496–1505. DOI: 10.1109/TIP.2007.896645 55

Chapra, S. C. (1980). *Applied Numerical Methods with MATLAB for Engineers and Scientists* (1 ed.). McGraw-Hill Higher Education, New York. 19

Chatterjee, S. and I. Chakrabarti (2011, November). Power efficient motion estimation algorithm and architecture based on pixel truncation. *IEEE Transactions on Consumer Electronics 57*(4), 1782–1790. DOI: 10.1109/TCE.2011.6131154 2

Chen, K. and D. a. Lorenz (2012, March). Image sequence interpolation based on optical flow, segmentation, and optimal control. *IEEE transactions on image processing : a publication of the IEEE Signal Processing Society 21*(3), 1020–30. DOI: 10.1109/TIP.2011.2179305 2, 62

Choi, C. and J. Jeong (2011, November). Low complexity weighted two-bit transforms based multiple candidate motion estimation. *IEEE Transactions on Consumer Electronics 57*(4), 1837–1842. DOI: 10.1109/TCE.2011.6131161 3

Chun, K. and J. Ra (1994). An improved block matching algorithm based on successive refinement of motion vector candidates. *Signal Processing: Image Communication 6*, 115–122. DOI: 10.1016/0923-5965(94)90010-8 5

Collins, D., A. Zijdenbos, V. Kollokian, J. Sled, N. Kabani, C. Holmes, and A. Evans (1998). Design and construction of a realistic digital brain phantom. *Medical Imaging, IEEE Transactions on 17*(3), 463–468. DOI: 10.1109/42.712135 50

Dawkins, M. S., R. Cain, and S. J. Roberts (2012, June). Optical flow, flock behaviour and chicken welfare. *Animal Behaviour 84*(1), 219–223. DOI: 10.1016/j.anbehav.2012.04.036 2

De Haan, G. and E. Bellers (1998, sep). Deinterlacing-an overview. *Proceedings of the IEEE 86*(9), 1839–1857. DOI: 10.1109/5.705528 60

Dufaux, F. and F. Moscheni (1995). Motion Estimation Techniques for Digital TV : A Review and a New Contribution. *Proceedings of the IEEE 83*(6), 858–876. DOI: 10.1109/5.387089 2

Eilers, P. (2004).   Parametric time warping.   *Analytical Chemistry 76*(2), 404–411. DOI: 10.1021/ac034800e 28

Fischer, B. and J. Modersitzki (2008). Ill-posed medicine—an introduction to image registration. *Inverse Problems 24*(3), 034008. DOI: 10.1088/0266-5611/24/3/034008 49

Frakes, D., C. Conrad, T. Healy, J. Monaco, M. Fogel, S. Sharma, M. Smith, and A. Yoganathan (2003). Application of an adaptive control grid interpolation technique to morphological vascular reconstruction. *Biomedical Engineering, IEEE Transactions on 50*(2), 197–206. DOI: 10.1109/TBME.2002.807651 43

Frakes, D., J. Monaco, and M. Smith (2001). Suppression of atmospheric turbulence in video using an adaptive control grid interpolation approach. In *Acoustics, Speech, and Signal Processing, 2001. Proceedings. (ICASSP '01). 2001 IEEE International Conference on*, Volume 3, pp. 1881 –1884 vol.3. DOI: 10.1109/ICASSP.2001.941311 21, 45

Frakes, D., K. Pekkan, L. Dasi, H. Kitajima, D. de Zelicourt, H. Leo, J. Carberry, K. Sundareswaran, H. Simon, and A. Yoganathan (2008). Modified control grid interpolation for the volumetric reconstruction of fluid flows. *Experiments in Fluids 45*, 987–997. DOI: 10.1007/s00348-008-0517-1 43, 44, 62, 65

Frakes, D., C. Zwart, and W. Singhose (2013, February). Extracting motion data from video using optical flow with physically-based constraints. *International Journal of Control, Automation and Systems 11*(1), in press. 43

Fujita, K., T. Hanada, Y. Kitazawa, and a. Kawabe (2012, March). A debris image tracking using optical flow algorithm. *Advances in Space Research 49*(5), 1007–1018. DOI: 10.1016/j.asr.2011.12.010 2

Giachetti, A. and N. Asuni (2008, September). Fast artifacts-free image interpolation. In *Proc. of the British Machine Vision Conf.*, Leeds, pp. 123–132. 55

Guo, K., X. Yang, H. Zha, W. Lin, and S. Yu (2012, February). Multiscale semilocal interpolation with antialiasing. *IEEE Transactions on Image Processing 21*(2), 615–625. DOI: 10.1109/TIP.2011.2165290 55

Han, J.-W., J.-H. Kim, S.-H. Cheon, J.-O. Kim, and S.-J. Ko (2010, February). A Novel Image Interpolation Method Using the Bilateral Filter. *IEEE Transactions on Consumer Electronics 56*(1), 175–181. DOI: 10.1109/TCE.2010.5439142 55

Horn, B. and B. Schunck (1981). Determining optical flow. *Artificial intelligence 17*(1), 185–203. DOI: 10.1016/0004-3702(81)90024-2 6

74 BIBLIOGRAPHY

Huang, C. and C. Hsu (1994). A new motion compensation method for image sequence coding using hierarchical grid interpolation. *Circuits and Systems for Video Technology, IEEE Transactions on 4*(1), 42–52. DOI: 10.1109/76.276171 10

Itakura, F. (1975, feb). Minimum prediction residual principle applied to speech recognition. *Acoustics, Speech and Signal Processing, IEEE Transactions on 23*(1), 67 – 72. DOI: 10.1109/TASSP.1975.1162641 27

Jain, J., H. Li, S. Cauley, C.-K. Koh, and V. Balakrishnan (2007, June). Numerically stable algorithms for inversion of block tridiagonal and banded matrices. Technical report, Purdue University, http://docs.lib.purdue.edu/ecetr/357/. 21

Jensen, K. and D. Anastassiou (1995). Subpixel edge localization and the interpolation of still images. *IEEE Transactions on Image Processing 4*(3), 285–295. DOI: 10.1109/83.366477 55

Jing, X., C. Zhu, and L. Chau (2003). Smooth constrained motion estimation for video coding. *Signal processing 83*(3), 677–680. DOI: 10.1016/S0165-1684(02)00482-6 2

Keogh, E. and M. Pazzani (2001, April). Derivative dynamic time warping. In *1st SIAM Int. Conf. on Data Mining (SDM-2001)*, Chicago, IL. 28

Kilthau, S., M. Drew, and T. Moller (2002). Full search content independent block matching based on the fast fourier transform. In *2002 International Conference on Image Processing.*, Volume 1, pp. I–669 – I–672 vol.1. DOI: 10.1109/ICIP.2002.1038113 3

Klein, A., J. Andersson, B. Ardekani, J. Ashburner, B. Avants, M. Chiang, G. Christensen, D. Collins, J. Gee, P. Hellier, et al. (2009). Evaluation of 14 nonlinear deformation algorithms applied to human brain mri registration. *Neuroimage 46*(3), 786. DOI: 10.1016/j.neuroimage.2008.12.037 49

Kovacs-Vajna, Z. (2000, nov). A fingerprint verification system based on triangular matching and dynamic time warping. *Pattern Analysis and Machine Intelligence, IEEE Transactions on 22*(11), 1266 – 1276. DOI: 10.1109/34.888711 27

Lee, H. Y., J. W. Park, T. M. Bae, S. U. Choi, and Y. H. Ha (2000, nov). Adaptive scan rate up-conversion system based on human visual characteristics. *Consumer Electronics, IEEE Transactions on 46*(4), 999 –1006. DOI: 10.1109/30.920453 60, 61

Lee, J.-H., J.-O. Kim, J.-W. Han, K.-S. Choi, and S.-J. Ko (2010, August). Edge-Oriented Two-Step Interpolation Based on Training Set. *IEEE Transactions on Consumer Electronics 56*(3), 1848–1855. DOI: 10.1109/TCE.2010.5606336 55

Lee, Y. J. and J. Yoon (2010, October). Nonlinear Image Upsampling Method Based on Radial Basis Function Interpolation. *IEEE Transactions on Image Processing 19*(10), 2682–2692. DOI: 10.1109/TIP.2010.2050108 55

Legrand, B., C. Chang, S. Ong, S.-Y. Neo, and N. Palanisamy (2008). Chromosome classification using dynamic time warping. *Pattern Recognition Letters 29*(3), 215 – 222. DOI: 10.1016/j.patrec.2007.09.017 27

Li, D., R. M. Mersereau, D. H. Frakes, and M. J. T. Smith (2005, September). A new method for suppressing optical turbulence in video. In *European Signal and Image Processing Conference*, Antalya, Turkey. 45

Li, X. and M. T. Orchard (2001). New edge-directed interpolation. *IEEE Transactions on Image Processing 10*(10), 1521–1527. DOI: 10.1109/83.951537 55

Li, Z., A. Martins da Silva, and J. a. P. S. Cunha (2002, June). Movement quantification in epileptic seizures: a new approach to video-EEG analysis. *IEEE transactions on bio-medical engineering 49*(6), 565–73. DOI: 10.1109/TBME.2002.1001971 2

Lim, H. and H. Park (2011). A symmetric motion estimation method for motion-compensated frame interpolation. *Image Processing, IEEE Transactions on 20*(12), 3653–3658. DOI: 10.1109/TIP.2011.2159232 62

Lin, S.-Y., C.-W. Su, and J.-S. Huang (2012, February). Expert system based parallel multi-1D block matching algorithm with implementation for motion estimation. *Expert Systems with Applications 39*(3), 3249–3256. DOI: 10.1016/j.eswa.2011.09.012 5

Little, J. and A. Verri (1989, March). Analysis of differential and matching methods for optical flow. In *Workshop on Visual Motion.*, pp. 173 –180. DOI: 10.1109/WVM.1989.47107 6

Liu, X., D. Zhao, R. Xiong, S. Ma, W. Gao, and H. Sun (2011, December). Image interpolation via regularized local linear regression. *IEEE Transactions on Image Processing 20*(12), 3455–3469. DOI: 10.1109/PCS.2010.5702437 55

Lucas, B. and T. Kanade (1981, April). An iterative image registration technique with an application to stereo vision. In *Proceedings of the 7th international joint conference on Artificial intelligence*, pp. 674–679. 6

Mallat, S. and G. Yu (2010). Super-resolution with sparse mixing estimators. *IEEE Transactions on Image Processing 19*(11), 2889–2900. DOI: 10.1109/TIP.2010.2049927 55, 56

Manjón, J. V., P. Coupe, A. Buades, V. Fonov, D. L. Collins, and M. Robles (2010, December). Non-local MRI upsampling. *Medical Image Analysis 14*(6), 784–792. DOI: 10.1016/j.media.2010.05.010 55

Mao, Y. and J. Gilles (2012). Non rigid geometric distortions correction-application to atmospheric turbulence stabilization. *Inverse Problems and Imaging 6*(3), 531–546. DOI: 10.3934/ipi.2012.6.531 45

Morse, B. S. and D. Schwartzwald (2001). Image magnification using level-set reconstruction. In *Proc. IEEE Computer Society Conf. Computer Vision and Pattern Recognition CVPR 2001*, Volume 1. DOI: 10.1109/CVPR.2001.990494 55

Nagel, H.-H. (1983). Displacement vectors derived from second-order intensity variations in image sequences. *Computer Vision, Graphics, and Image Processing 21*(1), 85 – 117. DOI: 10.1016/S0734-189X(83)80030-9 6

Nakaya, Y. and H. Harashima (1994). Motion compensation based on spatial transformations. *Circuits and Systems for Video Technology, IEEE Transactions on 4*(3), 339–356. DOI: 10.1109/76.305878 10

Oh, H.-S., Y. Kim, Y.-Y. Jung, A. Morales, and S.-J. Ko (2000). Spatio-temporal edge-based median filtering for deinterlacing. In *Int. Conf. on Consumer Electronics, 2000. ICCE. 2000 Digest of Technical Papers.*, pp. 52 –53. DOI: 10.1109/ICCE.2000.854493 60, 62

Park, S.-J., S.-M. Hong, H. Lee, S. Jin, and J. Jeong (2012). Histogram ordering model-based fast motion estimation. *IET Image Processing 6*(3), 238. DOI: 10.1049/iet-ipr.2010.0234 3

Park, S.-J. and J. Jeong (2010, November). Hybrid Image Upsampling Method in the Discrete Cosine Transform Domain. *IEEE Transactions on Consumer Electronics 56*(4), 2615–2622. DOI: 10.1109/TCE.2010.5681148 55

Penney, G. P., J. A. Schnabel, D. Rueckert, M. A. Viergever, and W. J. Niessen (2004). Registration-based interpolation. *IEEE Transactions on Medical Imaging 23*(7), 922–926. DOI: 10.1109/TMI.2004.828352 62

Purwar, R. K., N. Prakash, and N. Rajpal (2011, November). A matching criterion for motion compensation in the temporal coding of video signal. *Signal, Image and Video Processing 5*(5), 133–139. DOI: 10.1007/s11760-009-0149-9 3

Ramani, S., P. Thevenaz, and M. Unser (2010, February). Regularized Interpolation for Noisy Images. *IEEE Transactions on Medical Imaging 29*(2), 543–558. DOI: 10.1109/TMI.2009.2038576 55

Rath, T. and R. Manmatha (2003, june). Word image matching using dynamic time warping. In *Computer Vision and Pattern Recognition, 2003. Proceedings. 2003 IEEE Computer Society Conference on*, Volume 2, pp. II–521 – II–527 vol.2. 27

Rodriguez, A., J. R. Rabuñal, J. L. Pérez, and F. Martínez-Abella (2012, September). Optical Analysis of Strength Tests Based on Block-Matching Techniques. *Computer-Aided Civil and Infrastructure Engineering 27*(8), 573–593. DOI: 10.1111/j.1467-8667.2011.00743.x 2

Saha, A., J. Mukherjee, and S. Sural (2011, October). A neighborhood elimination approach for block matching in motion estimation. *Signal Processing: Image Communication 26*(8-9), 438–454. DOI: 10.1016/j.image.2011.06.002 5

Shabayek, A. E. R., C. Demonceaux, O. Morel, and D. Fofi (2011, August). Vision Based UAV Attitude Estimation: Progress and Insights. *Journal of Intelligent & Robotic Systems 65*(1-4), 295–308. DOI: 10.1007/s10846-011-9588-y 2

Shi, Z., W. A. C. Fernando, and A. Kondoz (2011). Adaptive Direction Search Algorithms based on Motion Correlation for Block Motion Estimation. *IEEE Transactions on Consumer Electronics 57*(3), 1354–1361. DOI: 10.1109/TCE.2011.6018894 5

Shindler, L., M. Moroni, and A. Cenedese (2012, May). Using optical flow equation for particle detection and velocity prediction in particle tracking. *Applied Mathematics and Computation 218*(17), 8684–8694. DOI: 10.1016/j.amc.2012.02.030 2

Song, Y. and A. Akoglu (2011, January). Bit-by-Bit Pipelined and Hybrid-Grained 2D Architecture for Motion Estimation of H.264/AVC. *Journal of Signal Processing Systems 68*(1), 49–62. DOI: 10.1007/s11265-010-0575-5 3, 5

Sullivan, G. and R. Baker (1991). Motion compensation for video compression using control grid interpolation. In *Acoustics, Speech, and Signal Processing, 1991. ICASSP-91., 1991 International Conference on*, pp. 2713–2716. IEEE. DOI: 10.1109/ICASSP.1991.150962 9, 10

Sun, D., S. Roth, and M. Black (2010). Secrets of optical flow estimation and their principles. In *Computer Vision and Pattern Recognition (CVPR), 2010 IEEE Conference on*, pp. 2432–2439. IEEE. DOI: 10.1109/CVPR.2010.5539939 44, 64

Szeliski, R. and J. Coughlan (1994, jun). Hierarchical spline-based image registration. In *Computer Vision and Pattern Recognition, 1994. Proceedings CVPR '94., 1994 IEEE Computer Society Conference on*, pp. 194 –201. DOI: 10.1109/CVPR.1994.323829 11

Temizel, A. (2007, October). Image resolution enhancement using wavelet domain hidden markov tree and coefficient sign estimation. In *Proc. IEEE Int. Conf. Image Processing (ICIP)*, Volume 5, pp. 381–384. DOI: 10.1109/ICIP.2007.4379845 56

Wang, Q., R. Ward, and H. Shi (2002). Isophote estimation by cubic-spline interpolation. In *Image Processing. 2002. Proceedings. 2002 International Conference on*, Volume 3, pp. III–401. IEEE. DOI: 10.1109/ICIP.2002.1038990 37

Wang, Q. and R. K. Ward (2007). A new orientation-adaptive interpolation method. *IEEE Transactions on Image Processing 16*(4), 889–900. DOI: 10.1109/TIP.2007.891794 56

Wiegand, T., G. J. Sullivan, S. Member, G. Bjø ntegaard, A. Luthra, and S. Member (2003). Overview of the H . 264 / AVC Video Coding Standard. *IEEE transactions on circuits and systems for video technology 13*(7), 560–576. DOI: 10.1109/TCSVT.2003.815165 2, 5

Wójcikowski, M., R. Żaglewski, and B. Pankiewicz (2011, January). FPGA-Based Real-Time Implementation of Detection Algorithm for Automatic Traffic Surveillance Sensor Network. *Journal of Signal Processing Systems 68*(1), 1–18. DOI: 10.1007/s11265-010-0569-3 2

Yamashita, Y., T. Harada, and Y. Kuniyoshi (2012). Causal Flow. *IEEE Transactions on Multimedia 14*(3), 619–629. DOI: 10.1109/TMM.2012.2191396 6

Zhang, X. and X. Wu (2008, June). Image interpolation by adaptive 2-D autoregressive modeling and soft-decision estimation. *IEEE Transactions on Image Processing 17*(6), 887–896. DOI: 10.1109/TIP.2008.924279 56

Zwart, C. and D. Frakes (2012, march). Soft adaptive gradient angle interpolation of grayscale images. In *Acoustics, Speech and Signal Processing (ICASSP), 2012 IEEE International Conference on*, pp. 845 –848. DOI: 10.1109/ICASSP.2012.6288016 56

Zwart, C., R. Pracht, and D. Frakes (2012). Improved motion estimation for restoring turbulence-distorted video. In *Society of Photo-Optical Instrumentation Engineers (SPIE) Conference Series*, Volume 8355, pp. 10. DOI: 10.1117/12.921125 43, 45

Zwart, C. M. and D. H. Frakes (2011, January). Biaxial control grid interpolation: Reducing isophote preservation to optical flow. In *Proc. IEEE Digital Signal Processing Workshop*, pp. 140–145. DOI: 10.1109/DSP-SPE.2011.5739201 56

Authors' Biographies

CHRISTINE M. ZWART

Christine M. Zwart received B.S.E. and M.S. degrees in bioengineering from Arizona State University where she is currently pursuing a Ph.D. in the same. She is a graduate researcher and doctoral candidate in the Image Processing Applications Lab working on image enlargement and enhancement algorithms with medical, defense, and consumer applications under the advisement of David H. Frakes. She is a National Science Foundation Graduate Research Fellow, a Science Foundation Arizona Graduate Research Fellow, a 2012 Arizona State University Faculty Women's Association Outstanding Graduate Student, and was engineering intern at The Boeing Company. After graduating she plans to work in medical imaging informatics at the Mayo Clinic.

DAVID H. FRAKES

David H. Frakes received B.S. and M.S. degrees in electrical engineering, an M.S. in mechanical engineering, and a Ph.D. degree in bioengineering, all from the Georgia Institute of Technology. In 2003, he co-founded 4-D Imaging, Inc., a small business that provides image and video processing solutions for the biomedical and military sectors. In 2008, he joined the faculty at Arizona State University (ASU) where he currently serves as a jointly appointed assistant professor in the School of Biological and Health Systems Engineering and the School of Electrical, Computer, and Energy Engineering. Professor Frakes was the ASU Centennial Professor of the Year in 2009, received the IEEE Outstanding University Faculty Award in 2011, and was awarded the National Science Foundation CAREER Award in 2012. He manages the Image Processing Applications Laboratory at ASU, which focuses on problems in image and video processing, machine vision, and fluid dynamics.

Printed in the United States
by Baker & Taylor Publisher Services